Revise AS Biology for OCR

Richard Fosbery, Jennifer Gregory and Ianto Stevens

Contents

Introduction – How to use this revision guide

This revision guide is for the OCR Biology AS course. Here is a plan of the modules you will study.

AS

Module name	Whole or half module?	Contents of module	Length of exam	Total number of marks in the exam (% of AS)	Distribution of marks in the exam
2801 Biology Foundation	whole	cell biology, biochemistry, enzymes, DNA/RNA, cell division, ecology	1 hour 30 minutes	90 (30%)	structured questions 75 marks extended questions 15 marks
2802 Health and Disease	whole	infectious and non-infectious diseases, diet, smoking-related diseases, exercise, immunity	1 hour 30 minutes	90 (30%)	structured questions 75 marks extended questions 15 marks
2803/01 Transport	half module	transport in mammals and flowering plants	1 hour	60 (20%)	structured questions 50 marks extended questions 10 marks
Practical Biology	half module	not covered in this revision guide – your school or college will organise this assessment which makes up 20% of the marks.			

It is likely that you will study 2801 and 2803/01 before 2802 since many of the topics in 2802 require knowledge of the other two modules. This is why we have put *Foundation* and *Transport* first. It is also possible that you will be taught the whole AS course in an integrated fashion rather than doing one module at a time.

When you are revising, have a copy of the specification. It is written as **learning outcomes** such as: *Candidates should be able to:*

(a) explain that enzymes are globular proteins which catalyse metabolic reactions.

This tells you exactly what you should learn. As you revise tick off each learning outcome when you are confident that you understand what it means and can write about it and explain it.

This book follows the sequence of learning outcomes in the specification. Each module is divided into a number of double page 'spreads', which end with some **quick check** questions. These are designed to test your recall and understanding of the topics you have just read. At the end of each section there are **end-of-module questions**, which resemble closely the types of questions you will find in the examination papers. **Answers** to selected quick check questions and all the end-of-module questions are on pages 86 to 93.

Good luck with your revision!

Module 2801: Biology Foundation

Cell structure

Viewing cells

To see and study cells, biologists use various types of microscope. These produce a magnified image that can be drawn or photographed. Two key concepts in microscopy are **resolution** and **magnification**.

Resolution is the ability to see detail. An image of a cell is formed when light, in a light microscope, or electrons in an electron microscope, are focused. Particles or membranes in cells can seen as separate objects if they are further apart than half the wavelength of light or the beam of electrons used. Electron microscopes have much greater resolution because electron beams have a shorter wavelength than light.

Magnification is the ratio between the size of an object and its image. It is calculated using the formula:

$$\text{Magnification} = \frac{\text{length of drawing or photograph}}{\text{length of object}}$$

When you do this sort of calculation, make sure you use the same unit for each measurement, e.g. millimetres (mm) or micrometres (µm).

This table compares light and electron microscopes.

feature	light microscope	electron microscope
wavelength	light – 400 nm	electron beam – 1.0 nm
resolution	200 nm	0.5 nm
maximum useful magnification	x 1500	x 250 000
image	natural colour (e.g. chlorophyll), coloured if dyes or stains are used	black and white – colour enhanced by computer
specimens	living or non-living	non-living
advantages	some living processes such as mitosis can be followed	very high resolution – can see plenty of cell detail

You should be able to use your GCSE knowledge to name all the parts of animal and plant cells that you can see in a **light** microscope.

Organelles

Animal and plant cells have structures called **organelles**. They are **eukaryotic** cells as their genetic material is contained within nuclei. This table gives details about the functions of organelles.

✓ *Quick check 1*

organelle	function(s)	key points
mitochondrion	aerobic respiration	highly folded internal membrane to give large surface area for enzymes
chloroplast	photosynthesis	grana made of stacks of membranes to give large surface area for chlorophyll and other pigments
nucleus	contains genetic information in DNA of chromosomes	separated from cytoplasm by nuclear envelope with pores for communication between nucleus and cytoplasm
nucleolus	production of ribosomes	darkly staining area in nucleus
ribosomes	amino acids assembled to make proteins	on rough endoplasmic reticulum (RER) or free in cytoplasm
rough endoplasmic reticulum (RER)	site for ribosomes, transports proteins to Golgi body	outer surface covered in ribosomes
Golgi body	modifies and packages proteins, makes lysosomes	flat sacs of membrane formed from endoplasmic reticulum (ER); give rise to vesicles or lysosomes
lysosomes	contain enzymes for destroying worn-out parts of cell and food particles	membrane keeps enzymes separate from rest of cell
centrioles	assembles the spindle to move chromosomes when nuclei divide	these replicate before division so that they are at each pole of the cell (see page 36)
smooth endoplasmic reticulum	makes triglycerides (fats), phospholipids and cholesterol	no ribosomes on the surface

Prokaryotes

Bacteria are **prokaryotes**. They have a much simpler cell structure than eukaryotes and have no nuclei.

✓ Quick check 2, 3

feature	prokaryote cell	eukaryote cell
cytoplasm	✓	✓
nucleus	✗	✓
nucleolus	✗	✓
cell wall	✓	✓
mitochondria	✗	✓
chloroplasts	✗	✓
Golgi body	✗	✓
endoplasmic reticulum	✗	✓
vacuole	✗	✓
typical diameter size (μm)	0.5–3.0	20–40

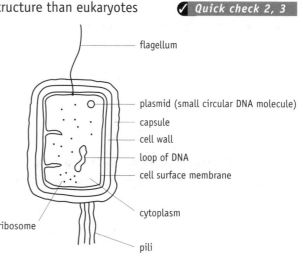

Fig. 1.1 A prokaryotic cell.

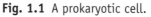

? Quick check questions

1 Explain the advantages of using the electron microscope to study cell structure.

2 Make a table to show the differences between plant cells, animal cells and prokaryotes.

3 Explain how the functions in plant and animal cells are divided between different organelles.

Tissues and organs

Some small organisms, such as *Stentor*, have bodies that are not divided into separate cells like ours. *Stentor* has **cilia** which it uses for moving and feeding. The cilia beat together in a pattern that you can see here to create a current of water. *Stentor* uses its cilia to filter small particles of food from the water.

Cilia are short cylindrical projections from the cell surface that beat back and forth.

- They are ideal for moving small organisms like *Stentor*.
- Animals use cilia for moving fluids past stationary cells.
- In the trachea, bronchi and bronchioles, cilia move a carpet of mucus.
- In fallopian tubes cilia move eggs from the ovary to the uterus.

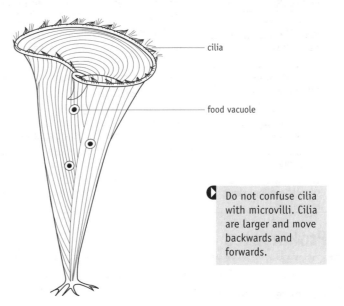

cilia

food vacuole

❰ Do not confuse cilia with microvilli. Cilia are larger and move backwards and forwards.

Fig. 2.1 *Stentor*, a freshwater protoctist with cilia that beat in a coordinated way to move it through water (× 35).

✓ *Quick check 1*

In animals, a layer of ciliated cells forms a **tissue** – a ciliated epithelium.

Tissue – a group of similar specialised cells in a many celled organism that carries out a specific function or several related functions.

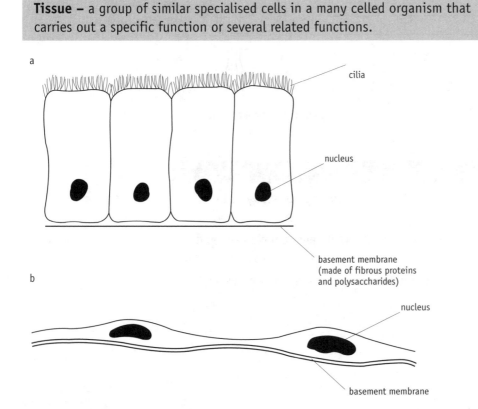

a

cilia

nucleus

basement membrane (made of fibrous proteins and polysaccharides)

b

nucleus

basement membrane

Fig. 2.2 Two tissues in the lungs (a) ciliated epithelium from the bronchus (see page 64) and (b) squamous epithelium from the alveoli.

The thin, flat cells that line the alveoli in the lungs form a tissue called squamous epithelium. Other examples of animal tissues are cartilage and bone (for support), muscle (for movement) and blood (for transport – see page 44).

✓ Quick check 2

Tissues are grouped into **organs**. Animals have many organs, such as lungs, hearts and kidneys. Roots, stems and leaves are plant organs.

> **Organ** – a group of different tissues forming a distinct structure and functioning together.

✓ Quick check 3

Plant tissues

Xylem and **phloem** are the transport tissues in plants. Xylem transports water and ions. Phloem transports sugars (mainly sucrose) and other compounds made by plants. Look at Fig. 2.3 to see where these tissues are in a plant.

✓ Quick check 4

▶ Make sure that you learn some examples of tissues and organs for both plants and animals. There are plenty on this spread to choose from.

▶ Note: when you are asked to draw plan drawings use lines to show the boundaries of the different tissues. Do not include *any* cells.

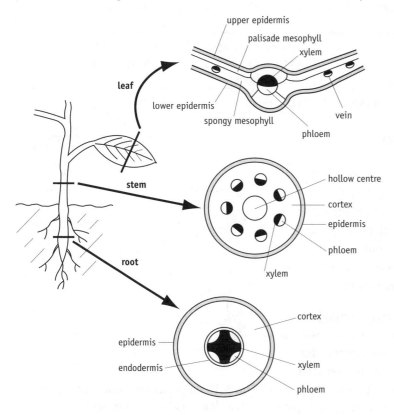

Fig. 2.3. Sections cut through the leaf, stem and root of a plant to show the distribution of tissues. Plan drawings like these are used to show tissues. They do not show any cells.

? *Quick check questions*

1 Describe three functions of cilia.

2 Name four different tissues in animals and state one function of each.

3 State the meanings of the terms **tissue** and **organ**.

4 Name two different plant tissues and state one function of each.

Biological molecules 1 – carbohydrates

In the next few spreads we look at some of the biological molecules that are made by living organisms. These are all based on the element carbon. The table shows the four different groups of **macromolecules**, the smaller molecules (sub-units or **monomers**) from which they are formed and the chemical elements that they contain.

macromolecules	examples	sub-unit molecule	chemical elements
carbohydrates	starch, glycogen, cellulose	glucose	C,H,O
proteins	haemoglobin, collagen, enzymes, e.g. amylase, protease, lipase	amino acids	C,H,O,N,S
lipids	triglycerides (fats and oils), phospholipids	glycerol, fatty acids, (and phosphate in phospholipids)	C,H,O (+ P for phospholipids)
nucleic acids	DNA and RNA (mRNA, tRNA and rRNA)	nucleotides	C,H,O,N,P

Glucose is a small carbohydrate

✓ Quick check 1

Carbohydrates are made of the elements carbon hydrogen and oxygen in the ratio $C_x(H_2O)_y$. Examples are glucose ($C_6H_{12}O_6$) and sucrose ($C_{12}H_{22}O_{11}$). Glucose is a source of energy in organisms and is built up into macromolecules. The two forms of glucose (shown in Fig. 3.1) differ only in the position of a hydroxyl group ($-OH$) about carbon atom 1. This simple difference makes a big difference to the macromolecules formed when they are **polymerised**.

Making and breaking a glycosidic bond

When cells make larger molecules from glucose monomers, the chemical bond that forms is a glycosidic bond. An oxygen atom acts as a 'bridge' between the glucose monomers. In maltose (see below), the oxygen links carbon 1 on one α glucose monomer with carbon 4 on the other. When glycosidic bonds form a molecule of water is eliminated (see Fig 3.2) in a **condensation** reaction. The addition of water breaks the bond and is known as **hydrolysis**.

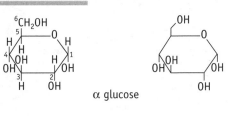

α glucose

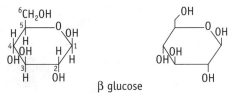

β glucose

Fig. 3.1 Alpha (α) and beta (β) glucose molecules in full and abbreviated forms. Notice that the carbon atoms are numbered 1 to 6 starting next to the oxygen atom and working clockwise.

Big carbohydrates

Plants polymerise α glucose to make two forms of **starch**:

- **Amylose** is a single, unbranched polymer which forms a helix. Iodine binds to the centre of the helix changing its colour from yellow to blue–black.

  This is why iodine solution changes colour when you use it to test for starch.
- **Amylopectin** is a branched chain. The branches are made by glycosidic bonds forming between carbon atoms 1 and 6 on glucose.

Animals polymerise α glucose to make **glycogen** which is like amylopectin, except that it branches much more often.

> ⬛ Make sure you can recognise and draw the simple forms of α and β glucose.

✓ Quick check 2

✓ Quick check 3

These three molecules are ideal for energy storage as they are insoluble, compact and provide large numbers of glucose molecules when needed for respiration. Glycogen and amylopectin have many 'ends' to which glucose can be added or taken off as required.

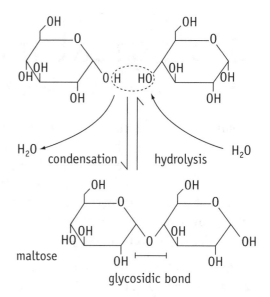

Plants make **cellulose** for cell walls. It is a polymer of β glucose. Alternate glucose molecules are turned through 180°. This means that cellulose forms straight chains with many projecting –OH groups that form hydrogen bonds along the molecule and with adjacent cellulose molecules. This makes cellulose very much stronger than starch and an ideal substance for cell walls, preventing plant cells from bursting when fully turgid.

Fig. 3.2 Glycosidic bonds form by condensation and are broken by hydrolysis.

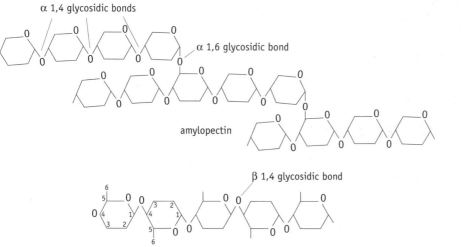

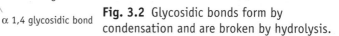

Fig. 3.3 Amylose, amylopectin and cellulose.

> If you are not sure about hydrogen bonds, see page 16. We will refer to hydrogen bonds several times throughout the book.

Testing for carbohydrates

- **Starch** gives a **blue–black** colour with iodine in potassium iodide solution.
- When boiled with Benedict's solution: **reducing sugars** (e.g. glucose) give a red precipitate and **non-reducing sugars** (e.g. sucrose) show no colour change at all.
- When boiled with dilute acid, cooled and neutralised with sodium hydrogencarbonate, **non-reducing** sugars will give a positive result with Benedict's solution because they have been hydrolysed to form reducing sugars.

✓ *Quick check 4*

? Quick check questions

1. Describe the difference between α and β glucose.
2. Describe how a glycosidic bond between two molecules of α glucose is formed and how it is broken.
3. Explain why starch and glycogen are good stores of energy.
4. Explain why cellulose is an ideal molecule for plant cell walls.

Biological molecules 2 – lipids

Lipids are large molecules with few oxygen atoms and many carbon and hydrogen atoms. This chemical composition makes them water repellent, or hydrophobic. They are less dense than water. Two groups of lipids are **triglycerides** (fats and oils) and **phospholipids**. These are made from sub-unit molecules – **glycerol** and **fatty acids**.

Glycerol is a three-carbon compound. Each carbon atom has a hydroxyl group (–OH group), which can react with a fatty acid to form an ester bond.

Fatty acids are long-chain hydrocarbon molecules. Each fatty acid has a carboxyl group at one end which can react with an –OH group on glycerol. Fatty acids vary as follows:

- length of the hydrocarbon chain (12 to 20 carbon atoms);

- number of double bonds in the chain.

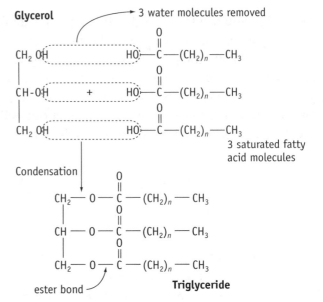

Fig. 4.1 A triglyceride is formed when three fatty acids form ester bonds with glycerol.

Saturated and unsaturated fatty acids

Saturated fatty acids have no double bonds in the hydrocarbon chain. Mono-unsaturated fatty acids have one double bond somewhere along the chain. Polyunsaturated fatty acids have two or more double bonds in the chain.

Triglycerides and phospholipids always have a mixture of different fatty acids. Triglycerides from fish and plants are oils because they contain more unsaturated than saturated fatty acids. An oil is a liquid at room temperature. Triglycerides from mammals are fats as they contain more saturated fatty acids. A fat is solid at room temperature.

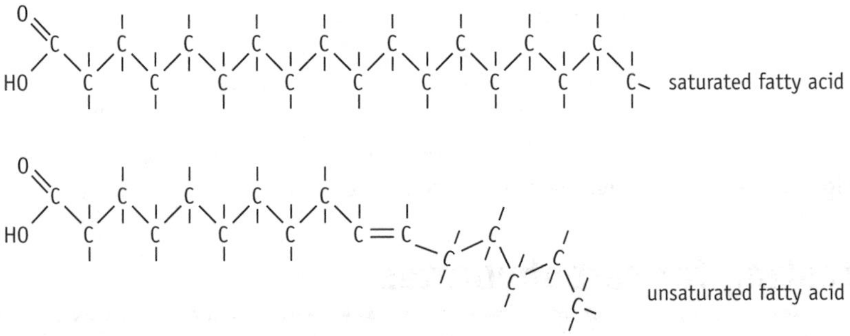

Fig. 4.2 This shows the hydrocarbon chains of a saturated and an unsaturated fatty acid. All the hydrogen atoms from the hydrocarbon chain have been omitted. The double bond causes the chain to bend.

✓ *Quick checks 1, 2*

Functions of triglycerides

Animals and plants use triglycerides for

- energy stores – when they are respired fats and oils release much energy (39 kJ g^{-1});

- thermal insulators – important to mammals and birds that live in cold climates (e.g. marine mammals, such as seals and whales);

- providing buoyancy – fats and oils are less dense than water;

- protecting internal organs, such as kidneys;
- providing an important source of water (especially for desert animals) when respired.

Phospholipids

Fig. 4.3 shows two ways in which we can show phospholipid molecules. The jagged lines on (a) and the straight lines on (b) are different ways of showing the hydrocarbon chains.

✓ *Quick check 3*

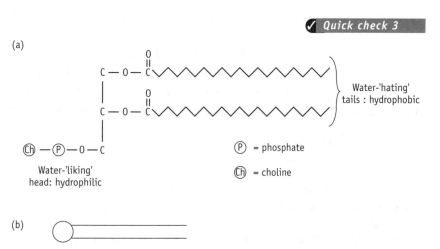

Attached to the phosphate are other groups (e.g. choline), which dissolve in water. The fatty acids, however, are hydrophobic, repelling water. Phospholipids therefore form stable bilayers when in water (Fig 4.4). These have hydrophobic cores and hydrophilic surfaces. The bilayer is an ideal structure for cell membranes as it creates a hydrophobic barrier between the parts of a cell or between a cell and its surroundings. Hydrophobic molecules such as fatty acids and oxygen cross bilayers very easily; small polar molecules, such as carbon dioxide and water, can also cross easily, but others cross more slowly or not at all. Cell surface membranes and the membranes surrounding organelles, such as mitochondria, have proteins to allow large polar molecules and ions across (see page 24).

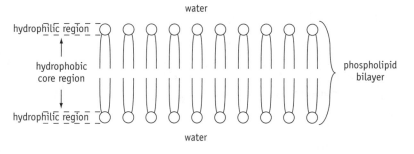

Fig. 4.3 (a) Phospholipids differ from triglycerides by having a phosphate group attached to glycerol in the place of a fatty acid. (b) shows a simple way of drawing a phospholipid.

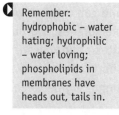 Remember: hydrophobic – water hating; hydrophilic – water loving; phospholipids in membranes have heads out, tails in.

✓ *Quick check 4*

Testing for lipids

Lipids are insoluble in water, but will dissolve in ethanol. When a solution of oil or fat in ethanol is poured into water, the ethanol helps tiny oil droplets to form and be dispersed throughout the water. The droplets scatter light that passes through, so you see a white cloudiness in the test-tube. This suspension is an emulsion and this is the **emulsion test** for lipids.

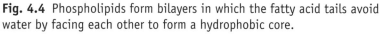

Fig. 4.4 Phospholipids form bilayers in which the fatty acid tails avoid water by facing each other to form a hydrophobic core.

❓ *Quick check questions*

1 Describe the structure of a triglyceride molecule.
2 Describe the ways in which triglycerides may differ from one another.
3 Explain how a phospholipid differs from a triglyceride.
4 Explain how phospholipids form bilayers.

Biological molecules 3 – proteins

Proteins are macromolecules made of chains of **amino acids**. Twenty different types of amino acids are used to make proteins. Amino acids differ by having different **residual groups** (R groups). A single chain of amino acids is a **polypeptide**. Some proteins consist of a single polypeptide; others are composed of two or more.

> Learn to draw the basic structure of all amino acids. You do not need to know the different R groups. Remember that each type of amino acid has its own specific R group.

Amino acids

Cells polymerise amino acids into proteins by forming peptide bonds (see page 32). A peptide bond forms between an amine group and a carboxylic acid group.

> ✔ *Quick check 1*

The properties of proteins depend on the residual groups (R groups) that project from the polypeptide chains. Some are charged (polar) and interact with water. Some are not charged and are hydrophobic and can interact with phospholipids within membranes. R groups determine the shape of active sites of enzymes (see page 18). Fig. 5.3 shows a short peptide, enkephalin, which is used for signalling between nerve cells in the brain.

(a) 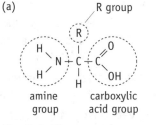 (b)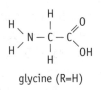

Fig. 5.1 The general structure of all types of amino acid (a) and the smallest, glycine (b).

> When drawing diagrams like Fig. 5.2 make sure you show that water is formed from an –OH group and an H.

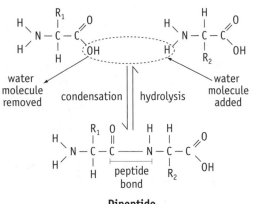

Dipeptide

Fig. 5.2 Peptide bonds form by condensation and are broken by hydrolysis. Two amino acids are joined by a peptide bond to form a dipeptide.

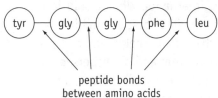

peptide bonds
between amino acids

Fig. 5.3 Enkephalin is a short peptide made of five amino acids. Polypeptides are made of ten or more amino acids. This diagram uses the three letter abbreviations for amino acids. tyr, tyrosine; gly, glycine; phe, phenylalanine; leu, leucine.

Organisation of a polypeptide

There are four levels of organisation of a protein, but we will deal with three here.

Primary structure of a polypeptide is its amino acid sequence. This is determined by the gene that codes for a polypeptide (see page 32).

Secondary structure is the folding of a polypeptide into one of three structures:

- α-helix, which is a right handed helix;
- β-pleated sheet, which is a flat sheet formed by a polypeptide that folds back on itself or links to adjacent polypeptides lying parallel to one another;
- no regular structure.

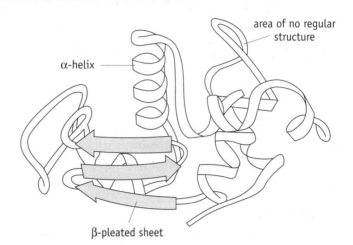

Fig. 5.4 The enzyme lysozyme is a polypeptide with 129 amino acids showing primary, secondary and tertiary structure. This 'ribbon' diagram shows the overall tertiary structure of lysozyme

Both the helix and the pleated sheet are stabilised by hydrogen bonds.

Tertiary structure is the further folding of a polypeptide to give a more complex three-dimensional shape. This shape is very precise and specific to the function of the polypeptide. Some polypeptides, such as the enzyme lysozyme, have areas of their tertiary structure composed of both α-helices and β-pleated sheets. Different parts of the polypeptide chain are close to one another and are stabilised by:

- hydrogen bonds
- disulphide bonds
- ionic bonds
- hydrophobic interactions.

Some of these bonds break when proteins are heated up or treated with acids and alkalis. When the bonds break the tertiary structure changes and the protein does not function. See page 22, which shows how this affects enzymes.

It is a good idea to mention the *shape* of proteins, such as cell surface receptors, enzymes and antibodies when explaining how these molecules combine with other molecules.

✓ *Quick check 2, 3*

Testing for proteins

Solutions of sodium hydroxide and copper sulphate are added to a test solution. A purple colour indicates the presence of a protein. A blue colour is a negative result. Sodium hydroxide breaks a protein into short peptides and the peptide bonds form coloured compounds with copper (II) ions in the copper sulphate.

? *Quick check questions*

1 Make an annotated diagram to show how a peptide bond forms between two amino acids.

2 Explain what is meant by primary, secondary and tertiary structure of a polypeptide.

3 Explain how the tertiary structure of a protein is stabilised.

Biological Molecules 4 – globular and fibrous proteins

Many proteins, such as enzymes and **haemoglobin**, are globular and are folded into complex 3D shapes. Fibrous proteins, such as **collagen**, do not form a complex shape and are insoluble. Both haemoglobin and collagen are composed of more than one polypeptide.

Haemoglobin

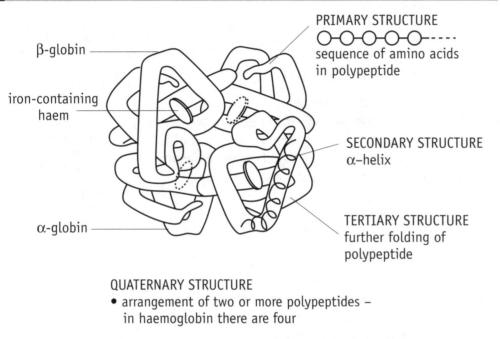

Fig. 6.1 Haemoglobin is a protein showing all four levels of organisation.

Red blood cells contain many molecules of haemoglobin, a protein which is specialised to transport oxygen. There are **four** polypeptides in each haemoglobin molecule. These polypeptides are called **alpha** and **beta globin** (α-globin and β-globin) and there are two of each. Each polypeptide has a tertiary structure stabilised by hydrophobic interactions in the centre, which maintain the overall shape.

In the middle of *each* polypeptide is a **haem group**, which is flat and circular with an atom of iron at the centre. Each haem group combines loosely with one oxygen molecule. This means that one molecule of haemoglobin can carry up to four molecules of oxygen.

Haemoglobin has quaternary structure

A protein has quaternary structure if it is made of two or more polypeptides. Haemoglobin has quaternary structure because it has four polypeptides, which fit together and are held in place by interactions between R groups on adjacent polypeptides: these interactions are hydrogen bonds and ionic bonds.

✓ *Quick check 1*

Knowing about the structure of haemoglobin will help you to explain how it works – see page 46.

✓ *Quick check 2*

Collagen is a fibrous protein

Collagen is found in skin, bone, cartilage, teeth, tendons, muscles, ligaments and the walls of blood vessels. It gives great toughness to these structures.

There are three identical polypeptide chains in a molecule of collagen and they are wound around each other to give a triple helix. Each polypeptide consists of about 1000 amino acids. In the primary structure every third amino acid is glycine, which has the smallest R group (one hydrogen atom) of all the amino acids.The sequences of the polypeptides are staggered so that glycine is always found at every position along the triple helix. This allows the three polypeptides to pack closely together to form many hydrogen bonds along their whole length. The many hydrogen bonds give the molecule its great strength.

Collagen does not show secondary, tertiary and quaternary structure in the same way as globular proteins, such as haemoglobin. The triple helix is a left-handed helix (an alpha helix has a right-hand turn) and there is no further folding to give a complex three-dimensional tertiary shape. Adjacent molecules of collagen form covalent bonds between R groups. These link many such molecules into fibres. You can see in Fig. 6.2 that the molecules are arranged so they overlap without any lines of weakness where the collagen fibre might break if pulled very hard. The many cross links and the hydrogen bonds within the fibre give it great strength.

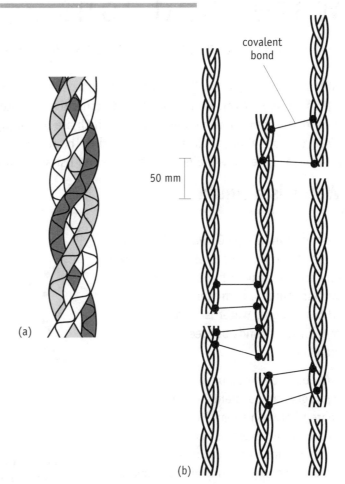

Fig. 6.2 (a) Collagen: three polypeptides are wound very tightly round each other to form a triple helix. (b) Covalent bonds link the helices together into a strong fibre.

✓ **Quick check 3, 4**

❓ *Quick check questions*

1 Each red blood cell may contain 280 million molecules of haemoglobin. What is the maximum number of oxygen molecules that a red cell can carry?

2 Explain why haemoglobin has quaternary structure.

3 Explain how the structure of collagen makes it an ideal substance for ligaments and tendons.

4 Make a table to show how the structure and functions of haemoglobin differ from those of collagen.

◖ Collagen is a protein and cellulose is a carbohydrate – but both are very tough. You should be able to compare them in terms of structure and function – see page 9 for details of cellulose.

Water and ions

Life originated in water. The bodies of animals are about 60–70% water. Plant tissues are about 90% water. Many animals and plants live in water. Sea water, freshwater and body fluids contain many ions, some of which are of biological importance. You will discover more about their importance during the rest of the AS course and during the A2 course. Here we explain what they are and list some of their functions.

Water is a liquid

From its formula (H_2O) water should be a gas at the temperatures we experience on Earth. A heavier molecule with a similar formula, hydrogen sulphide (H_2S), is a gas. Water is a liquid because of hydrogen bonds between water molecules.

The bonds between oxygen and hydrogen are covalent, with electrons shared between them. However, the oxygen atom exerts an attraction for the electrons in the covalent bonds making the oxygen slightly negatively charged (δ^-) and the hydrogen slightly positively charged (δ^+). The symbol δ (delta) indicates a slight charge. The attraction between δ^+ and δ^- is a hydrogen bond and each water molecule may form up to four of these to make a cluster. In water, the clusters break and reform all the time.

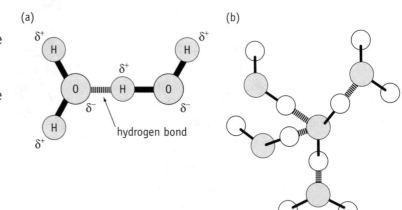

Fig. 7.1 (a) A water molecule showing the uneven distribution of charge and (b) a cluster of water molecules with hydrogen bonds between them.

> Hydrogen bonds are important in stabilising proteins and DNA.

> ✓ *Quick check 1*

Water as an environment

Water provides an environment that:
- has a fairly stable temperature;
- provides support;
- provides some oxygen (solubility of oxygen is low in water);
- is transparent so light can penetrate for plants.

Water molecules are cohesive

Hydrogen bonds make water molecules 'sticky'. This stickiness is known as **cohesion** and is responsible for many of the properties of water that are important to living things.

This table shows the key features of water:
- as an environment for organisms;

- as a constituent of living organisms.

property	key points	roles of water
good solvent for charged and uncharged substances	water molecules are attracted to ions and polar (charged) molecules, e.g. glucose	transport in blood, xylem and phloem
specific heat capacity	4200 kJ are necessary to increase the temperature of 1 kg of water by $1^{\circ}C$. The thermal energy absorbed is used to break hydrogen bonds	helps to prevent changes in body temperature and in water surrounding aquatic organisms
latent heat of vaporisation	much thermal energy is used to cause water molecules to change to water vapour	coolant – water is used efficiently, as a small amount of water absorbs much thermal energy
density	ice is less dense than water so it floats on water	ice insulates water beneath allowing organisms to survive
high cohesion	hydrogen bonds 'stick' water molecules together	helps draw up water in xylem, gives surface tension

✓ *Quick check 2, 3*

Ions

An ion is a charged particle. Some ions, such as calcium, are positively charged (known as **cations**); others, such as chloride, are negatively charged (**anions**). They are found in the soil, in water and are important constituents of living things. This table shows some important functions of ions in animals and in plants (in italics).

ion	functions
calcium Ca^{2+}	used by muscles and synapses (gaps between nerve cells) to transmit nerve impulses; used to make bones and teeth; *used to make calcium pectate that holds cell walls of neighbouring plant cells together*
sodium Na^{+}	transmission of electrical impulses in nerve cells; maintains osmotic balance
potassium K^{+}	transmission of electrical impulses in nerve cells; *opening and closing of guard cells in stomata*
magnesium Mg^{2+}	used to make bones and teeth; *used to make chlorophyll*
chloride Cl^{-}	used to make stomach acid (HCl); maintains osmotic balance
nitrate NO_3^{-}	*used to make amino acids and nucleotides*
phosphate PO_4^{3-}	used to make phospholipids, ATP, nucleotides, tooth enamel and bone; *used to make phospholipids, ATP, nucleotides*
iron Fe^{2+}	part of haemoglobin (page 14) which transports oxygen

You may be asked to state one function of each of these ions. You may learn more about these ions in the A2 course.

❓ *Quick check questions*

✓ *Quick check 4*

1 Explain why water is a liquid at $40^{\circ}C$ and not a gas like hydrogen sulphide.

2 List the functions of water in living organisms.

3 What are the advantages and disadvantages of living in water?

4 What is an ion?

Enzymes

Many chemical reactions occur in living organisms. Most of these reactions would occur very slowly, if we did not have special catalysts.

Enzymes are biological catalysts

Enzymes are catalysts made of protein. Enzymes convert substrates into products by having active sites where reactions occur. Some enzymes speed up reactions where molecules are broken down; others catalyse reactions where large molecules are built up from smaller ones.

This table summarises the properties of enzymes.

protein properties	catalyst properties
globular proteins specific shape formed of tertiary structure provide active site where reaction occurs influenced by temperature and pH denatured by extremes of pH and temperatures above optimum	increase rates of reaction specific to a reaction remain unchanged at end of reaction lower activation energy

> Enzymes catalyse hydrolysis and condensation reactions. (See pages 9, 10 and 12 for examples).

Activation energy

Energy is stored in the bonds that hold atoms together in a molecule. When substrates react, bonds are broken and new bonds form. When there is no enzyme, reactions do not occur readily because the substrates (or substrate) do not have enough energy to be converted to a product. The extra energy that needs to be given is the **activation energy**. Enzymes decrease activation energy by providing an **active site** where reactions occur more easily than elsewhere.

Enzymes allow covalent bonds in substrate molecules to be broken and energy released. Often in exergonic reactions, energy is transferred to ATP. This happens in respiration.

> Reactions will occur faster if heated. Since the structure of most proteins is permanently changed at temperatures greater than 40°C, this is not an option for living organisms.

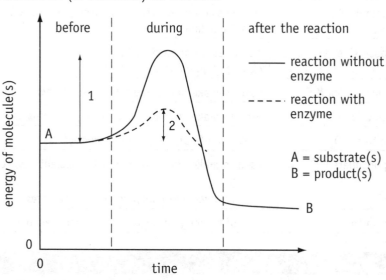

1 activation energy without enzyme
2 activation energy with enzyme

Fig. 8.1 An enzyme lowers activation energy for this exergonic reaction in which energy is released from a substrate.

Action of enzymes

In a solution, enzyme and substrate molecules are constantly moving and frequently collide. A substrate molecule may fit into the enzyme active site. It is held in the active site for a brief moment, forming an **enzyme–substrate complex**. The reaction occurs and the product or products are formed which then leave the active site. The active site is now free to receive another substrate molecule.

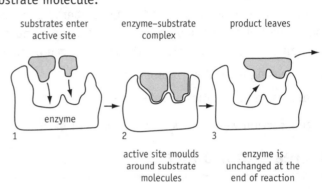

> The active site is part of the enzyme, not the substrate.

In Fig. 8.2 you can see the enzyme changes shape to mould itself around the substrate molecule. This is called **induced fit**. As this happens R groups on the polypeptide forming the active site are brought into a position which combines with the substrate, forming temporary bonds with it. Two substrate molecules may be brought close together in such a way that a reaction occurs, where outside the enzyme it would occur only slowly. The bonds within a single substrate may be put under strain so that it is broken down.

Fig. 8.2 Enzyme molecules go through asa change in shape as they bind to their substreates.

✓ *Quick check 1,2,3*

Enzyme specificity

Active sites are specific for one type of molecule. Some examples of enzyme specificity:

> See page 12 to remind yourself about R groups.

- amylase breaks down glycosidic bonds in starch to form maltose;
- catalase breaks down hydrogen peroxide into water and oxygen. It has four polypeptides each with an active site;
- subtilisin is a protease which breaks peptide bonds between any pair of amino acids;
- trypsin is a protease that only breaks peptide bonds next to arginine and lysine.

> This is all to do with *shape* again. Amylase will not break down cellulose because only starch has the right shape to fit in the active site.

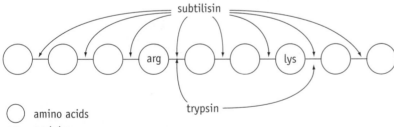

amino acids
arg = arginine
lys = lysine

Fig. 8.3

❓ *Quick check questions*

1 Define the terms **enzyme**, **substrate**, **product** and **active site**.

2 Explain how an enzyme increases the rate of a reaction.

3 Make a diagram similar to Fig. 8.2 to show a substrate molecule breaking apart to form two product molecules.

Experiments with enzymes

When an enzyme and its substrate are mixed together, a reaction begins. Substrate molecules collide with the enzyme and bind to its active site. Product molecules are formed. As the reaction proceeds, the number of substrate molecules decreases and the number of product molecules increases. The number of enzyme molecules stays constant. The speed, or rate, of the reaction can be followed by measuring either

- increasing quantities of product, or
- decreasing quantities of substrate.

In this section we will look at an example of each.

Increasing product

Fig. 9.1 shows some apparatus that may be used to measure the activity of the enzyme catalase in breaking down hydrogen peroxide:

$$2H_2O_2 \rightarrow O_2 + 2H_2O$$

The rate of the reaction is determined by measuring how much oxygen is collected at intervals of time and using this formula:

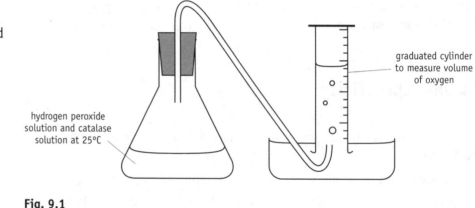

hydrogen peroxide solution and catalase solution at 25°C

graduated cylinder to measure volume of oxygen

Fig. 9.1

$$\text{rate of reaction} = \frac{\text{volume of oxygen produced}}{\text{time}}$$

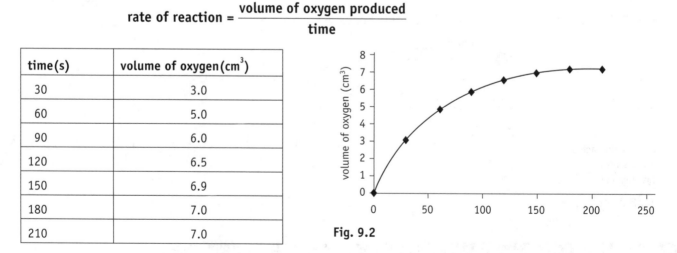

time (s)	volume of oxygen (cm^3)
30	3.0
60	5.0
90	6.0
120	6.5
150	6.9
180	7.0
210	7.0

Fig. 9.2

Fig. 9.2 shows these results plotted on a graph.

As the reaction proceeds, less oxygen is produced as there is less substrate available. The rate of reaction is quickest at the beginning when there is a high concentration of substrate. Later, substrate concentration becomes a limiting factor so the reaction slows down. Eventually all the substrate is used up so the reaction stops.

Decreasing substrate

Plants and animals break down starch to maltose using the enzyme amylase.

$$\text{starch} + \text{water} \xrightarrow{\text{amylase}} \textbf{many molecules of maltose}$$

The apparatus in Fig. 9.3 is used to follow the breakdown of starch in this reaction. Samples are taken from the reaction mixture and tested with iodine solution. At the beginning, there is plenty of starch in the reaction mixture and the colour with iodine is dark blue. Later, most of the starch has been hydrolysed and the colour is very light blue. When all the starch is broken down the colour is yellow. The colorimeter measures the absorbance of samples taken from the reaction mixture. A dark colour with iodine (indicating much starch present) gives a high value for absorbance.

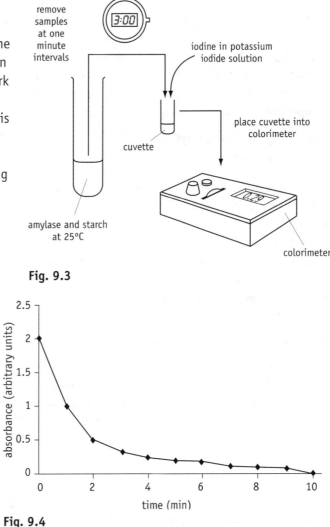

Fig. 9.3

time (min)	absorbance (arbitrary units)
0	2.00
1	1.01
2	0.55
3	0.29
4	0.23
5	0.19
6	0.16
7	0.11
8	0.07
9	0.05
10	0.00

Fig. 9.4

A graph of the results (Fig. 9.4) shows the absorbance readings decreasing rapidly at the start and more slowly later on. The reasons are the same as before. At the start there are many molecules of starch and all the active sites of the enzyme are filled. Later as the concentration of starch falls, there are many active sites unfilled and the rate of reaction falls.

Remember that in both examples described here results are being taken at intervals of time from the same reaction mixture.

✔ *Quick check 1,2,3*

❓ Quick check questions

1 Describe what happens to the substrate concentration during the course of an enzyme catalysed reaction.

2 Explain why the reaction catalysed by amylase occurs quickly at the start and then slows down.

3 What do you think would happen to the reactions catalysed by catalase and amylase if they were carried out: (a) at 35°C and not 25°C; (b) with higher concentrations of enzyme?

Factors affecting enzyme activity

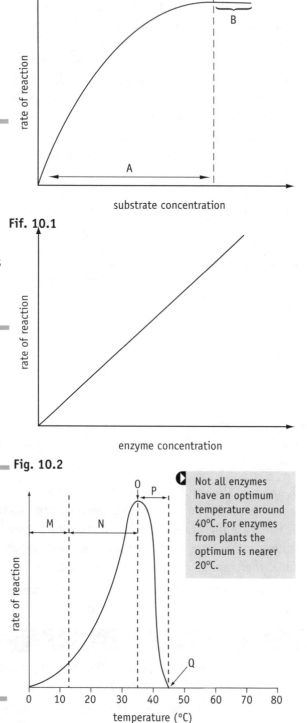

What happens to the activity of enzymes if various factors are changed? The first part of this section deals with four such factors: **substrate concentration**, **enzyme concentration**, **temperature** and **pH**.

Substrate concentration

Each point plotted on Fig. 10.1 is the rate of reaction with different starting concentrations of substrate, but the same concentration of enzyme has been used each time. In **A**, the rate increases as the substrate concentration increases. Substrate concentration is the limiting factor. In **B**, the rate does not increase any further as the enzyme concentration is limiting. All the active sites are filled.

Fif. 10.1

Enzyme concentration

The rate of reaction increases as the concentration of enzyme increases. When there is plenty of substrate the only limiting factor is the enzyme concentration. With more enzyme molecules there are more active sites available.

Fig. 10.2

Temperature

At **M** there is a slow reaction because molecules of enzyme and substrate have little kinetic energy and rarely collide. At **N**, there is more kinetic energy and collisions occur more frequently. **O** is the optimum temperature where the rate is fastest. **P** is where the enzyme molecules begin to lose their tertiary structure. Bonds that hold the polypeptides in specific shapes are broken and the shape of the active site changes. Substrate molecules no longer fit. At **Q**, the enzyme is denatured and loses all activity.

Not all enzymes have an optimum temperature around 40°C. For enzymes from plants the optimum is nearer 20°C.

Fig. 10.3

pH

Most enzymes work over a narrow range of pH. Fig. 10.4 shows the optimum pH for three different enzymes. The optimum pH for enzymes that work in cells, such as catalase, is about 7.0. Pepsin works in the stomach where hydrochloric acid is secreted. This explains its low optimum pH. Trypsin works in the small intestine, which has a pH of about 8.0.

✓ *Quick check 1*

When the pH changes from the optimum:

- the shape of the enzyme changes;
- the affinity of the substrate for the active site decreases.

Many of the bonds that hold an enzyme's tertiary structure together are bonds between positively and negatively charged R groups. When the pH changes there is a change in the concentration of hydrogen ions which cancels out the charge on these R groups so the tertiary structure changes, which means that active sites lose their specific shape and substrate molecules do not fit.

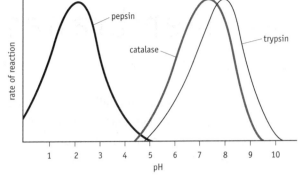

Fig. 10.4

Enzyme inhibitors

Inhibitors prevent enzymes from working. Some have a temporary effect and are reversible. There are two types of reversible inhibitor, **competitive** and **non-competitive**.

Competitive inhibitors are molecules that enter the active site. They have a shape similar to that of the substrate so they fit into the active site. When this happens, the substrate cannot enter so the enzyme does not catalyse the reaction. Inhibitor and substrate molecules cannot occupy the active site at the same time. Inhibitor molecules constantly move in and out of the active site, so it is possible to reduce the effect of the inhibitor by adding more substrate.

Non-competitive inhibitors do not occupy the active site, but fit into another site on the enzyme molecule. This changes the shape of the active site so that the substrate can no longer fit. If the concentration of this type of inhibitor is high enough, all enzymes may be inhibited and the reaction slows to nothing. Increasing the concentration of the substrate has no effect on this type of inhibition.

✓ *Quick check 2*

Competitive inhibitors fit into the active site; non-competitive inhibitors fit elsewhere on the enzyme.

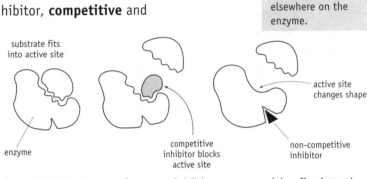

Fig. 10.5 Two types of enzyme inhibitor – competitive fits into the active site, non-competitive fits into another site on the enzyme.

✓ *Quick check 3*

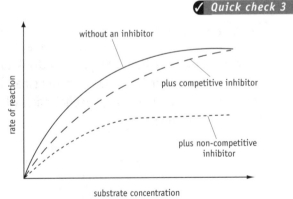

Fig. 10.6 Increasing the substrate concentration can overcome the effect of competitive inhibitors, but does not have the same effect with non-competitive inhibitors.

Quick check questions

1. Explain the effect of temperature on enzyme activity.
2. Explain why enzymes often only function efficiently over a narrow range of pH.
3. Describe two ways in which enzymes may be inhibited.

Cell membranes

Cell surface membranes surround all cells and control what enters and leaves. Membranes divide up the cytoplasm of eukaryotic cells into separate compartments. In this section we look at the structure of membranes and some of their functions both at the cell surface and within the cell.

Membranes are very thin

Membranes are visible in the electron microscope at magnifications of x 100 000 as two dark lines separated by a clear space. The distance across the membrane is between 7 and 10 nm. All membranes are made of two layers of phospholipid, known as a bilayer, and protein. The polar heads of the phospholipids are hydrophilic and are attracted to water. This is why they face towards the cytoplasm and towards the exterior of the cell. Both of these areas are dominated by water. The hydrocarbon tails of the two layers are hydrophobic and are held together by weak hydrophobic bonds. Proteins are scattered about the membrane and transmembrane proteins pass right through from one side to the other. This structure is called a fluid mosaic. The phospholipid bilayer is split open in Fig. 11.1 to show the proteins that pass right through the membrane. Carbohydrates are attached to protein and lipid and face the outside of the cell.

> Most of the organelles you have to recognise are made of membrane – see page 5.

> Phospholipids form a bilayer: 'heads out', 'tails in' – see page 11.

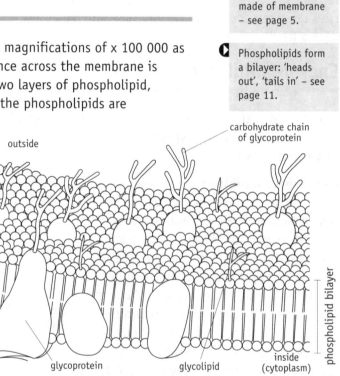

Fig. 11.1

Why fluid mosaic?

- **Fluid**: phospholipids are fluid and you can think of a membrane as a thin layer of oil.
- **Mosaic**: a mosaic is a picture made of many small pieces of tile. Proteins are like the pieces of tile surrounded by phospholipids, which are like the cement holding everything together.

✓ *Quick check 1, 2*

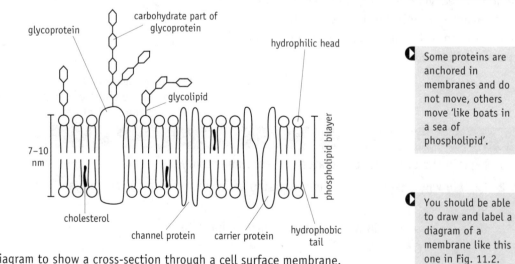

Fig. 11.2 A simple diagram to show a cross-section through a cell surface membrane.

> Some proteins are anchored in membranes and do not move, others move 'like boats in a sea of phospholipid'.

> You should be able to draw and label a diagram of a membrane like this one in Fig. 11.2.

component of cell membrane	functions
phospholipids	<ul><li>form a bilayer to act as a barrier between cytoplasm and cell exterior</li><li>fluid so components can move within the membrane</li><li>permeable to non-polar molecules, such as oxygen and fatty acids</li><li>permeable to small polar molecules, such as ethanol, water and carbon dioxide</li><li>impermeable to ions and large polar molecules, such as sugars and amino acids</li></ul>
cholesterol	<ul><li>stabilises the phospholipid bilayer by binding to polar heads and non-polar tails of phospholipids</li><li>controls fluidity by preventing phospholipids solidifying at low temperatures and becoming too fluid at high temperatures.</li></ul>
proteins	<ul><li>transmembrane proteins act as channels and carriers for e.g. ions and glucose</li><li>receptors for hormones</li></ul>
carbohydrates	<ul><li>attached to lipids to form glycolipids and attached to proteins to form glycoproteins</li><li>only found on exterior surface of cell membranes and act as receptors and aid in recognition of cells.</li></ul>

Functions of membranes

✓ Quick check 3

Membranes are *partially permeable* because some substances pass through but others do not. The permeability of a membrane is determined by the phospholipids and the proteins. Membranes are selective about what passes through. Cell surface membranes:

- act as a barrier to many water-soluble substances;
- keep many large molecules such as enzymes within the cell;
- are permeable to small molecules such as water, oxygen and carbon dioxide;
- are permeable to selected molecules such as glucose and ions;
- permit movement of substances by endocytosis and exocytosis;
- permit recognition by other cells, such as those of the immune system;
- provide receptors for signalling molecules such as hormones;
- are often extended into microvilli to provide a large surface area for the absorption of substances by animal cells.

❶ Do not confuse microvilli with cilia – see page 6.

Membranes within cells:

- divide the cell into compartments where functions can occur more efficiently;
- isolate potentially harmful enzymes in lysosomes;
- provide a large surface for pigments, such as chlorophyll, involved in photosynthesis in chloroplasts;
- provide large surface for holding the enzymes for forming ATP in mitochondria and chloroplasts;
- form small vacuoles which transport molecules between parts of the cell, e.g. proteins from RER to Golgi body.

✓ Quick check 4

❓ Quick check questions

1 Explain why membranes are described as *fluid mosaic*.
2 Make a diagram to show the structure of a fluid mosaic membrane.
3 Describe the functions of phospholipids, cholesterol, proteins and carbohydrates in membranes.
4 Describe the distribution of membranes within typical animal and plant cells.

Exchanges across membranes

Substances are exchanged between cells and their surroundings, across cell surface membranes. Cells obtain all their requirements in this way and their waste substances and some of their products pass in the opposite direction. Molecules move across membranes by:

- diffusion, including osmosis;
- facilitated diffusion;
- active transport;
- bulk transport (**endocytosis** and **exocytosis**).

Diffusion

Diffusion is a passive process that does not require any energy on the part of cells. Molecules like oxygen and carbon dioxide diffuse across membranes so long as a concentration gradient exists. Both molecules are small and uncharged, so they diffuse through the phospholipid bilayer very easily. In the lungs concentration gradients are maintained by:

- breathing, which refreshes air in the alveoli;
- the continuous flow of deoxygenated blood into the lungs from the heart.

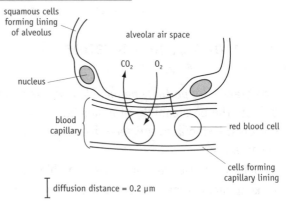

Fig. 12.1

Osmosis

Water molecules are small enough to pass between phospholipid molecules and they also pass through pore proteins. The movement of water into and out of cells is influenced by a variety of factors:

- how much water is present in the cytoplasm and in the exterior environment;
- the concentration of solutes, such as ions and sugars, on either side of the cell surface membrane;
- in plants, the pressure exerted on cell contents by the cell wall.

> Always explain the movement of water in and out of cells using water potential gradients. There are more examples on pages 51 and 55.

Water potential is the tendency for water to move from one place to another and is determined by the factors listed above. Solutions with high water potentials have few solute molecules; solutions with low water potentials have many dissolved solute molecules. Water moves from a solution with a high water potential to one with a lower water potential. The diffusion of water through a partially permeable membrane down a water potential gradient is **osmosis**.

Facilitated diffusion

Many molecules that cells require are too large to pass between phospholipid molecules. They may also be charged and therefore unable to pass through the hydrophobic region in the centre of the bilayer. Protein molecules exist in membranes to help (or *facilitate*) the diffusion of these substances. ✓ *Quick check 1, 2*

- **Channel proteins** are transmembrane proteins that form tunnels, or pores, through the bilayer for water-soluble molecules. Some channels are open all the time, others open when triggered by the presence of a chemical such as a hormone. The lining of the pore is formed by hydrophilic R groups of the protein that allow water and polar substances to pass through easily.

- **Carrier proteins** change shape to help move molecules into or out of the cell. Molecules bind to the protein which stimulates the protein to change its overall shape, so allowing the molecules to diffuse through the membrane.

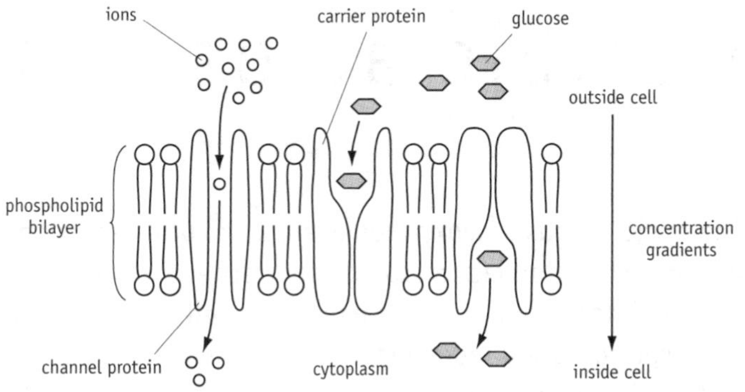

Fig. 12.2 Facilitated diffusion through channel and carrier proteins.

Active transport

Some substances required by cells are in a lower concentration outside the cell than inside. Cells cannot obtain these substances by diffusion. Root hair cells absorb nutrients, such as potassium ions, from the water in the soil. Carrier proteins similar to those used in facilitated diffusion move ions across membranes but *against* the concentration gradient. Root hair cells use energy from respiration to change the shape of carrier proteins so that they can pump molecules or ions into cells. Active transport is also used to pump molecules and ions out of cells.

> Carrier proteins use energy from the cell in active

Bulk transport

- **Exocytosis.** Substances packaged by Golgi bodies are delivered to the cell surface in small vacuoles (sometimes called vesicles). These vacuoles fuse with the membrane to push out their contents.

- **Endocytosis.** Some cells take up large molecules, such as proteins, and much larger solid objects, such as bacteria, by enclosing them inside small vacuoles formed by membrane. ✓ *Quick check 3,4*

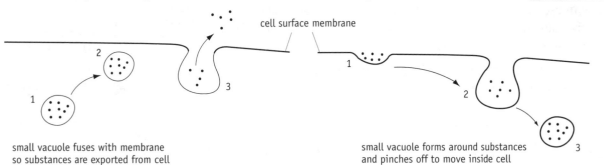

small vacuole fuses with membrane so substances are exported from cell

Exocytosis

small vacuole forms around substances and pinches off to move inside cell

Endocytosis

Fig. 12.3 Cells use exocytosis to secrete substances and endocytosis to take in large molecules and particles.

Quick check questions

1 Define the terms **diffusion**, **facilitated diffusion**, **osmosis** and **active transport**.

2 Explain why carbon dioxide diffuses through phospholipid bilayers, but glucose does not.

3 Describe how exocytosis occurs at a cell surface.

4 Explain why (i) alveoli are lined by a squamous epithelium, (ii) root hair cells have mitochondria and (iii) channel proteins are lined by amino acids with hydrophilic R groups.

Nucleic acids – DNA and RNA

The nucleic acids – deoxyribonucleic acid (DNA) and ribonucleic acid (RNA) – make up the cell's information storage and retrieval system. DNA is a long-term store of genetic information that is passed from cell to cell during growth and from parent to offspring during reproduction. RNA molecules are smaller and do not last as long as DNA. They have a number of different functions in information retrieval when proteins are synthesised in cells.

Nucleic acids are polynucleotides

DNA and RNA are macromolecules made of chains of nucleotides. Each nucleotide consists of:

- pentose sugar (with five carbon atoms);
- phosphate;
- nitrogen-containing base.

There are two different types of base. The larger bases are the purines adenine and guanine. They have double rings of carbon and nitrogen atoms. The smaller bases are the pyrimidines thymine, cytosine and uracil. These have a single ring of carbon and nitrogen atoms. The bases are often known by the letters A, G, T, C and U. A, G, C and T are in DNA; the bases A, G, C and U are in RNA.

> DNA and RNA are made of nucleotides - *not* amino acids.

A nucleotide is shown in diagrammatic fashion in Fig. 13.1. DNA nucleotides have the sugar deoxyribose, RNA nucleotides have the sugar ribose. Both sugars are pentoses – they have five carbon atoms. The numbers refer to the carbon atoms. (Note that this is a diagram and many of the groups – such as −OH – have been left out.)

> *Remember*: uracil replaces thymine in RNA.

The nucleotides are attached to each other by covalent bonds between phosphate and sugar molecules to form a 'backbone'.

> ✓ *Quick check 1*

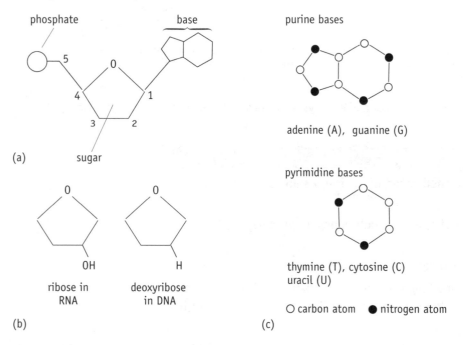

Fig. 13.1 (a) A nucleotide. (b) Ribose and deoxyribose. (c) Simplified structure of purine and pyrimidine bases.

DNA is a double helix

Fig. 13.2 shows the double helix of DNA. It is made of two polynucleotides held together by hydrogen bonds. The bases are shown as blocks projecting inwards, the dashed lines represent hydrogen bonds.

The bases are always arranged so that a purine is opposite a pyrimidine. Hydrogen bonding always occurs between the bases so that adenine is always paired with thymine and guanine is always paired with cytosine. There are two hydrogen bonds between A and T and three between C and G. The bases are complementary in size and shape so that only the pairings A–T and C–G fit into the space between the sugar–phosphate backbone of DNA.

▶ The rules of base pairing (A–T and C–G) always apply. Look out for more examples in the next few sections

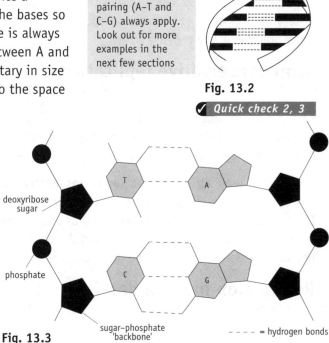

Fig. 13.2

✓ *Quick check 2, 3*

DNA is well suited for the long-term storage of genetic information as it

- is a very stable molecule so the information stored in DNA is kept for a long time;
- is a large molecule that stores huge amounts of information;
- each of the two polynucleotides acts as a template for synthesis of a new polynucleotide during DNA replication (see page 30).

deoxyribose sugar

phosphate

sugar–phosphate 'backbone'

- - - - = hydrogen bonds

Fig. 13.3

The three forms of RNA

RNA differs from DNA in that the molecule is a single, shorter polynucleotide. RNA contains the sugar ribose, not deoxyribose and the base uracil, not thymine. The table shows the three forms of RNA and their functions.

type of RNA	structure of polynucleotide	function
messenger RNA (mRNA)	variable length, no base pairing	transfers genetic information from DNA to ribosomes, after which it is broken down
transfer RNA (tRNA)	clover-leaf structure with some base pairing	carries amino acids to ribosomes
ribosomal RNA (rRNA)	folded and attached to proteins to make ribosomes	provides site for assembly of amino acids to make proteins

tRNA molecules are used in protein synthesis to identify amino acids and transfer them to ribosomes. They have attachment sites for amino acids, ribosomes and mRNA.

✓ *Quick check 4*

❓ *Quick check questions*

1 Define the following terms: **nucleic acid**, **nucleotide** and **polynucleotide**.

2 Describe the structure of a DNA molecule.

3 Explain what is meant by **base pairing** in DNA.

4 Make a table to show how the structure of DNA differs from the structure of RNA.

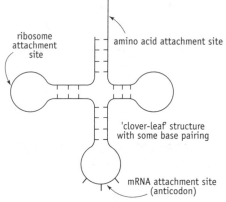

ribosome attachment site

amino acid attachment site

'clover-leaf' structure with some base pairing

mRNA attachment site (anticodon)

Fig. 13.4 A tRNA molecule.

Replication of DNA

DNA is the only molecule that can be copied. There are enzymes in cells that make copies of DNA, but they need some DNA to start with. This copying process, or replication, is very precise. There are few, if any copying errors.

Semi-conservative replication

Each polynucleotide acts as a template for making a new polynucleotide. This means that a new molecule of DNA consists of an older polynucleotide and a recently made one.

This type of replication is called **semi-conservative** as half of the 'parent' DNA molecule is passed to the 'daughter' molecule. In this way the 'parent' DNA is *conserved* as one polynucleotide forms half of one daughter molecule and the other polynucleotide forms part of the other one.

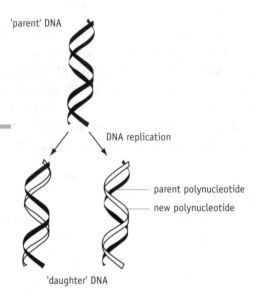

Fig. 14.1

Replication in more detail

1 The double helix unwinds and the DNA 'unzips' as hydrogen bonds between the polynucleotides are broken.

2 Existing polynucleotides act as templates for assembly of nucleotides.

3 Free nucleotides which have been made in the cytoplasm move towards the exposed bases of DNA.

4 Base pairing occurs between free nucleotides and exposed bases. A matches T and C matches G. Hydrogen bonds form between complementary bases.

5 The enzyme DNA polymerase forms covalent bonds between the free nucleotides, which have become attached to each template.

6 Two 'daughter' DNA molecules form separate double helices.

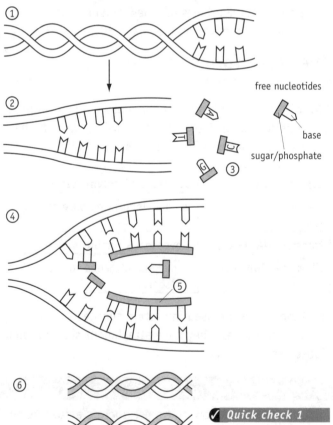

✓ Quick check 1

Fig. 14.2 What happens to a small section of DNA when it is replicated.

Experimental evidence for replication

Evidence for the semi-conservative method described above comes from the experiment shown in Fig. 14.3. Bacteria replicate their DNA before they divide.

When they are grown in a solution with a heavy isotope of nitrogen, they make bases and then nucleotides which are heavier than normal. These 'heavy' nucleotides are then used to make a 'heavy' form of DNA. DNA extracted from bacteria that have grown with the two types of nitrogen are spun in a centrifuge and compared. The DNA can be seen under ultraviolet light. DNA with the 'heavy' form of nitrogen settles at a lower depth in the centrifuge tube than DNA with the light form.

Bacteria that had been growing in the 'heavy' medium for many generations were transferred to a medium containing the 'light' form of nitrogen. They were given enough time to replicate their DNA and divide and then a sample of their DNA was spun in a centrifuge (Generation 1). Further samples were taken (Generations 2 and 3). DNA produced by the bacteria after the transfer (Generation 2 in Fig. 14.3) have DNA which is between the 'heavy' form and the 'light' form. This supports the idea that new DNA molecules are composed of an existing polynucleotide and a newly synthesised one. Fig. 14.3 shows how the experiment was carried out and the results.

> Check that you can explain the results of this experiment by using what is shown in Fig. 14.1.

> ✓ **Quick check 2**

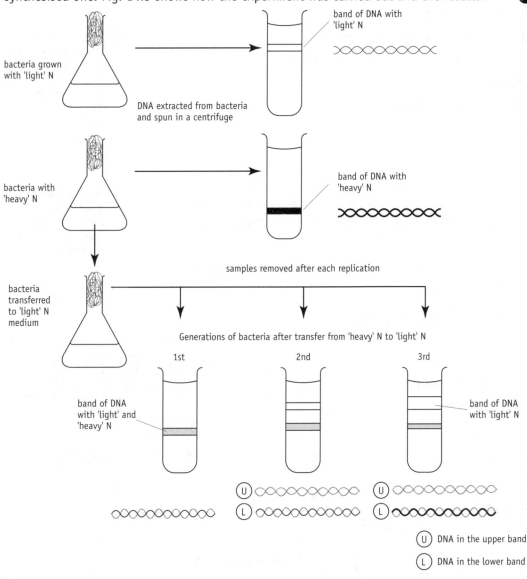

Fig. 14.3

? Quick check questions

1 Describe what happens when DNA is replicated.

2 Outline the experimental evidence for semi-conservative replication of DNA.

Protein synthesis

There is a huge amount of information stored in the DNA of our chromosomes. Each cell has a full complement of chromosomes with all the genetic information needed to make the whole body. However, individual cells need to retrieve only a small quantity of this information. For example, cells in our salivary glands use the gene that codes for the enzyme amylase. Liver cells use the gene that codes for catalase. Each gene is a code for the primary structure of a polypeptide; the sequence of bases in DNA determines the sequence of amino acids in the polypeptide, which is made by ribosomes in the cytoplasm. Cells have thousands of ribosomes that use short-lived transcripts of the genes in the form of messenger RNA. The process of protein synthesis occurs in four stages.

- Transcription of DNA to make messenger RNA (mRNA).
- Movement of mRNA from the nucleus to the cytoplasm.
- Amino acid activation.
- Translation of mRNA to make a polypeptide.

> ◖ The primary structure of a protein is its sequence of amino acids – see page 13.

Transcription

DNA 'unzips' along the length of the gene so that the enzyme RNA polymerase can match free RNA nucleotides to form a molecule that is complementary to the coding template. This follows the rules of base pairing. The other polynucleotide takes no part in the process.

Each group of three bases in mRNA, or triplet, codes for a specific amino acid. These triplets are called **codons**.

> ◖ mRNA is made by assembling nucleotides when a transcript of a gene is needed.

> ◖ Note that U replaces T in RNA.

Movement of mRNA to ribosomes

When transcription is finished, the messenger RNA molecule moves from the chromosome to a ribosome in the cytoplasm. In a eukaryotic cell, mRNA moves from the nucleus where transcription occurs, through nuclear pores, to ribosomes.

Amino acid activation

Enzymes attach amino acids to their specific tRNA molecule. This needs energy supplied by ATP. The **anticodon** is a triplet of bases forming part of the tRNA molecule and it is complementary to a codon.

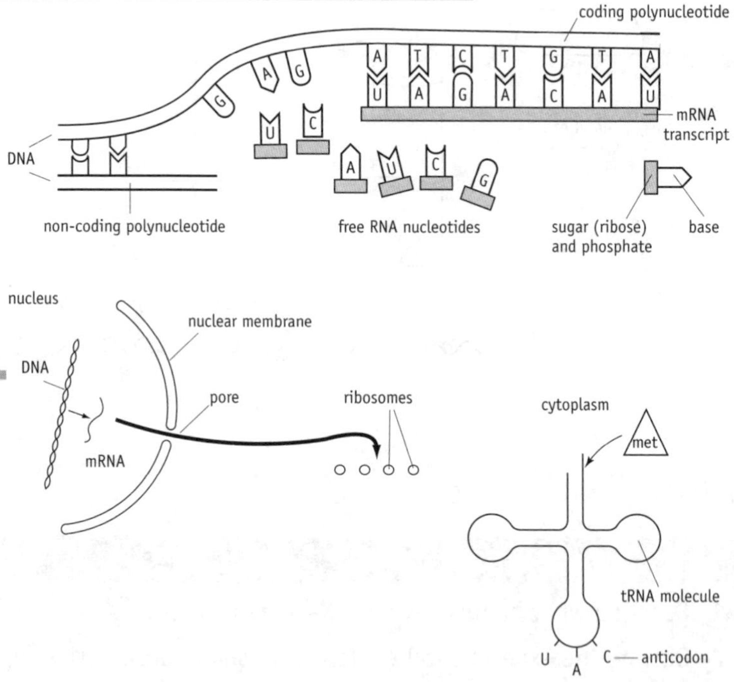

Fig. 15.1

Translation

You do not have to remember any of the triplet codes, but by following the rules of base pairing you should be able to work out the mRNA codons and tRNA anticodons from information you are given.

Fig. 15.2 shows the process of translation.

1. The mRNA molecule binds to a ribosome and translation begins. The first codon is the start codon (AUG), which codes for the amino acid methionine. The anticodon (UAC) on the tRNA molecule forms base pairs with the codon on mRNA.

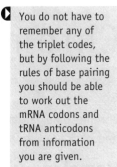

2. Another tRNA molecule (here with serine) occupies the second site in the ribosome. A peptide bond forms between methionine and serine.

3. The ribosome moves one codon along the mRNA. The tRNA for methionine leaves and another tRNA (here with alanine) occupies the vacant position. Notice how tRNA molecules with anticodons that match the mRNA codons ensure that the genetic message is 'read' correctly.

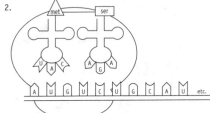

4. As the ribosome moves along the mRNA molecule more amino acids are added to the end of the polypeptide. This carries on until the ribosome reaches a stop codon (UAA, UAG or UGA). There are no tRNA molecules for these codons and so the polypeptide breaks lose from the ribosome and translation is complete.

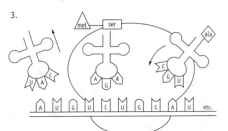

There are twenty different amino acids used to make proteins and 64 different arrangements of the four bases. This means that there are several codons and several types of tRNA for many amino acids.

There are more codons than there are amino acids. There are four RNA codons for glycine, GGG, GGU, GGA and GGC.

DNA	RNA codon	tRNA anticodon	amino acid
TAC	AUG	UAC	methionine (start)
AGA	UCU	AGA	serine
CGT	GCA	CGU	alanine
CCC	GGG	CCC	glycine
CAG	GUC	CAG	valine
ATT ATC ACT	UAA UAG UGA	none	none

Fig. 15.2

✓ Quick check 1,2,3

❓ Quick check questions

1 Describe the roles of DNA, RNA polymerase, mRNA, ribosomes, and tRNA in the synthesis of a polypeptide.

2 The following are codons: GGA (glycine), UCG (serine) and AAG (lysine). What are the tRNA anticodons?

3 Explain why there are 64 different codons, but only 61 different anticodons.

Chromosomes and genes

Chromosomes are made of DNA and protein. Proteins are in chromosomes to support and package the DNA, which is the genetic material.

Homologous chromosomes

Fig. 16.1 is a drawing made from a photograph of all the chromosomes from a human cell. On the right the same chromosomes are arranged into homologous pairs.

When photographed like this, chromosomes are double structures. Replication has occurred so that each chromosome consists of two molecules of DNA that are packaged tightly to make two sister chromatids. As replication is very precise, the two sister chromatids are genetically identical.

> ◗ You inherited your chromosomes from your parents. In each homologous pair there is a maternal and a paternal chromosome.

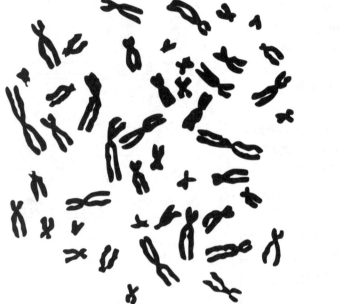

Fig. 16.1

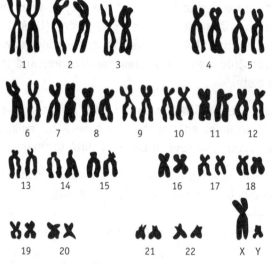

Fig. 16.2

Chromosomes from Fig. 16.1 have been matched and put into pairs to give the arrangement on the right, which is called a karyotype. Pairs of chromosomes are homologous. The chromosomes in each pair are the same size, they have the same shape, the centromere is always in the same place and they have the same genes. (Note that the sex chromosomes, X and Y, have only one small region that is homologous.)

✓ **Quick check 1**

Human females have two X chromosomes. Fig. 16.3 shows the positions of six different genes on the X chromosome and the features that they control. Genes are parts of chromosomes that determine the sequence of amino acids that make up polypeptides.

Here the gene for factor VIII determines the sequence of amino acids that make up this important blood clotting protein. Some people inherit a non-functioning gene for factor VIII, do not make the blood clotting protein and suffer from the genetic disease haemophilia. Their blood takes a long time to clot and they are in danger of bleeding internally (haemorrhaging). Haemophilia is treated by injection of factor

VIII collected from blood donations. Some people treated in this way became infected by human immunodeficiency virus (HIV) or hepatitis. Factor VIII is now made by genetically modified hamster cells and, as a result, is much safer. Other drugs and medicines are made in a similar way. Human insulin is made by genetically modified bacteria and has been available since 1982.

Genetic engineering

As DNA in all organisms has the same structure, it is possible to join together pieces of DNA from different sources. Insulin is a hormone that controls blood sugar. It is a protein. People with diabetes often cannot make insulin and need daily injections of the hormone. The gene for human insulin was inserted into bacteria so that large quantities of human insulin could be manufactured rather than extracting it from dead animals.

Here are the steps that were taken to modify bacteria and use them to produce insulin.

Preparation of insulin gene

- mRNA for human insulin extracted from pancreas cells;
- reverse transcriptase enzyme uses mRNA as a template to make complementary DNA (cDNA);
- cDNA has a single sequence of nucleotides (GGG) added to each end to make 'sticky ends'.

Preparation of a vector to carry the human gene into a bacterium

- plasmid (small circular form of DNA) cut open with a restriction enzyme;
- cut plasmid has a single sequence of nucleotides (CCC) added to each end to make 'sticky ends';
- cDNA for insulin and plasmids are mixed so that 'sticky ends' form base pairs;
- ligase enzyme links sugar phosphate backbones of cDNA to plasmid.

Formation of genetically modified bacteria

- plasmids mixed with bacteria;
- bacteria take up plasmids and multiply to form a clone;
- genetically modified bacteria transcribe and translate human gene to make human insulin.

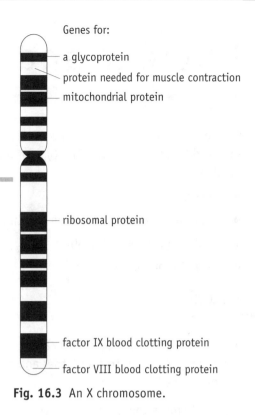

Genes for:
- a glycoprotein
- protein needed for muscle contraction
- mitochondrial protein
- ribosomal protein
- factor IX blood clotting protein
- factor VIII blood clotting protein

Fig. 16.3 An X chromosome.

✓ *Quick check 2*

? Quick check questions

1 Explain what is meant by **homologous pairs of chromosomes**.
2 Describe how the following are used in genetic engineering: **reverse transcriptase**, **restriction enzyme** and **plasmid**.

Nuclear and cell division

Cells grow to a certain size and then divide into two. The nucleus divides first followed by the cytoplasm. The type of nuclear division involved is **mitosis** and it is involved in:

- growth;
- replacement of cells, e.g. skin and lining of the gut;
- repair, e.g. in wound healing;
- asexual reproduction.

Mitosis

All the cells produced by mitosis are genetically identical to each other and to the parent cell.

Fig. 17.1 shows the mitotic cell cycle. A cell can only divide after it has replicated its DNA in the 'S' (for synthesis) stage that occurs during interphase. DNA synthesis and mitosis require energy. During the 'G' (for gap) stages, cells build up their energy reserves and make new membrane and organelles.

During mitosis, the duplicated chromosomes separate and move to opposite ends of the cell. A cell may then divide into two to give two daughter cells. As a result of mitosis:

- the number of chromosomes in a nucleus stays the same;
- the genetic information passed to the daughter cells is identical;
- two new nuclei are formed;
- no genetic variation occurs.

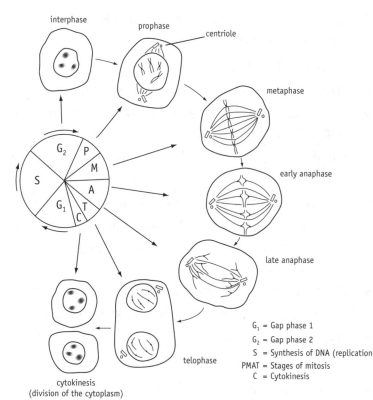

Fig. 17.1

G_1 = Gap phase 1
G_2 = Gap phase 2
S = Synthesis of DNA (replication
PMAT = Stages of mitosis
C = Cytokinesis

Stages of mitosis

1 Prophase: the DNA in chromosomes is packaged. Chromosomes shorten and thicken. This makes it easy for the cell to move chromosomes around.

The chromosomes are now condensed and through the light microscope you can see that each chromosome has two **chromatids**. The nuclear envelope begins to break up into small pieces and disperse throughout the cell. There is now no boundary between nucleus and cytoplasm.

2 Metaphase: chromosomes come to the middle of the cell. A **spindle** made from protein is organised by the **centrioles**. Chromosomes are attached to the spindle at the **centromere**.

3 Anaphase: chromatids break apart at the centromere and are pulled by the spindle towards the poles.

4 Telophase: nuclear envelopes reform around each group of chromosomes at either end of the cell. The chromatids uncoil.

The chromosome number of the new nuclei is exactly the same as the chromosome number of the original nucleus. The new cells formed in mitosis have exactly the same genetic information as each other although they may end up looking quite different because they may express different genes. Telophase is followed by an interphase in which DNA may be replicated if the cell is going to divide again.

✓ *Quick check 1,2*

Uncontrolled mitosis leads to cancers

Mitosis is usually under strict control. Cancer-causing agents (or carcinogens), such as X-rays, ultraviolet light and tar from cigarettes bring about changes to the genes that control cell division. When these genes change, or mutate, cells can start to divide in an uncontrolled fashion. A mass of unspecialised cells forms a tumour, which may grow so large as to block organs, such as the bronchi in the lungs, or to obstruct the flow of blood. Parts of the tumour may break off and be carried in the blood to other parts of the body where they cause secondary cancers in other organs. This makes it difficult to treat cancer successfully.

✓ *Quick check 3*

Life cycles

From generation to generation, the chromosome number in the body cells of species that reproduce sexually remains constant. This number of chromosomes in the body cells is the **diploid** number. The human diploid number is 46. When gametes (eggs and sperm) are produced a different type of nuclear division occurs – meiosis.

During meiosis the chromosome number is halved so that human gametes have 23 chromosomes – one of each type. The number of chromosomes in gametes is the **haploid** number. For humans, the haploid number is 23. This ensures that at fertilisation the diploid number of chromosomes is restored.

◗ Remember that in mitosis the chromosome number stays constant; in meiosis it is halved.

✓ *Quick check 4*

❓ *Quick check questions*

1 Describe the behaviour of one chromosome during a mitotic cell cycle.

2 Explain why replication must always occur before mitosis.

3 Describe how a tumour forms.

4 Explain the significance of meiosis in life cycles.

Energy and ecosystems

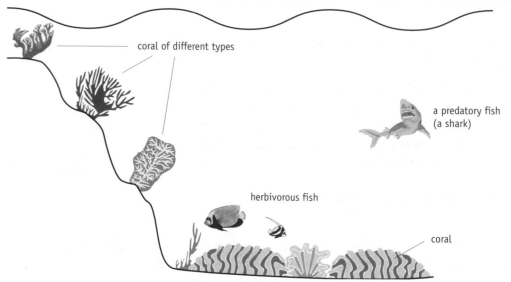

coral of different types

a predatory fish
(a shark)

herbivorous fish

coral

Fig. 18.1

This picture shows part of a coral reef with herbivorous fish grazing on algae that grow over the surface of the rock.

Many organisms live on or around coral reefs forming a **community**. The reef is their **habitat**. Individual species, like the fish, occupy a particular **niche** on the reef. All the fish of the same species on this reef form a **population**. The reef with its community of organisms and the physical environment form an **ecosystem**. Energy flows through ecosystems. **Producers**, such as algal films on the reef, convert light energy from the sun into chemical energy. **Primary consumers**, such as herbivorous fish, feed on the algae, becoming food for **secondary consumers**, such as predatory fish. This simple feeding relationship is a **food chain**:

> Make sure that you can define these terms and use them correctly when writing about ecosystems.

$$\text{algae} \rightarrow \text{herbivorous fish} \rightarrow \text{predatory fish}$$

In ecosystems, such as coral reefs, the feeding relationships are complex since each animal eats a variety of different foods and is eaten by a number of different predators. **Food webs** show these different feeding relationships.

Ecosystem definitions

Community All the organisms living in one easily defined area.

Habitat The place where an organism lives.

Population All the organisms belonging to the same species living in the same area at the same time. Males and females in a population can interbreed.

Niche The role of a species in a community. The role refers to its position in the food chain and how it interacts with the environment and with other species.

Ecosystem A self-contained community, all the physical features that influence the community and the interactions between organisms and their environment.

Producer An organism that converts simple inorganic compounds (e.g. carbon dioxide and water and ions) into complex organic compounds. Most use light to provide the energy to drive the reactions involved.

Consumer Organism that gains energy from complex organic matter, e.g. herbivores, carnivores, omnivores, detritivores, decomposers, parasites.

Food chain shows energy flows from one organism to another in a simple sequence.

Food web shows energy flow through many organisms in an ecosystem.

Trophic level Each feeding level in a food chain.

Energy flow

✓ *Quick check 1*

Only a small part of the energy entering a trophic level becomes available to the next trophic level. The percentage varies according to the food chain concerned and the efficiency with which energy is transferred. Not much energy reaches animals at the top of food chains (e.g. tigers). This explains why they are rare.

Energy is lost in food chains because animals:

- never eat all the available food;
- cannot digest all the food they eat;
- use energy in their respiration so they can move, hunt, chew, reproduce, etc.;
- lose heat energy to their surroundings;
- lose energy in urine and faeces; this energy may pass to decomposers.

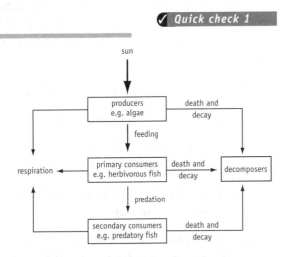

Fig. 18.2 Energy flow on a coral reef.

Nitrogen cycle

All organisms need nitrogen as it is a component of essential compounds, such as amino acids, proteins, nucleotides, nucleic acids (DNA and RNA). Nitrogen in the atmosphere (dinitrogen, N_2) is inert and few organisms can gain their nitrogen from this source. Some bacteria, such as *Rhizobium*, are able to fix nitrogen from the air by reducing it to ammonium ions (NH_4^+). This process of **nitrogen fixation** requires much energy. Most nitrogen fixation occurs in swollen nodules on the roots of leguminous plants, such as peas and beans. The plants provide sugars, from their photosynthesis, to the bacteria so that they have the energy to split the nitrogen molecule (N_2) and combine it with hydrogen. *Rhizobium* and the host plants use ammonium ions to make amino acids. Both then use amino acids to make proteins.

Nitrogen combined with another element, such as hydrogen, is called **fixed nitrogen**. Most organisms use forms of fixed nitrogen rather than use dinitrogen. Fig. 18.3 shows the main transformations that occur to nitrogen in plants, animals and microorganisms.

> There is more to the nitrogen cycle than this, but you need to know the roles of these bacteria in cycling nitrogen.

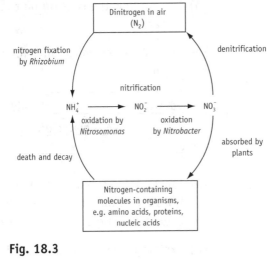

Fig. 18.3

✓ *Quick check 2, 3*

Quick check questions

1 Describe the flow of energy in a named ecosystem.

2 Explain how nitrogen in the protein of a dead animal is made available to plants.

3 Explain the roles of *Rhizobium*, *Nitrosomonas* and *Nitrobacter* in the nitrogen cycle.

End-of-module questions

These questions will show you the types of questions that are likely to be on the examination paper for the Foundation module. There is also some advice on examination technique. You will find the answers to these questions on pages 86 and 91.

1 Fig. 19.1 is an electron micrograph of part of an epithelial cell from a lung.

> Do not be surprised to find something unfamiliar like this – look carefully and try to recall what you remember of cell structure.

(a) (i) Name the structures **A** to **D**. [4]

  (ii) Explain what is meant by an organelle. [2]

  (iii) Calculate the actual length of the cell in micrometres. [2]

(b) State **three** ways in which the structure of a palisade cell from a leaf differs from that of the epithelial cell shown in Fig. 19.1. [3]

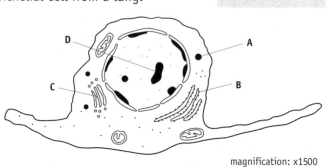

magnification: x1500

Fig. 19.1

(c) Explain why palisade mesophyll is a tissue, while the lung is an organ. [3]

(d) State **three** structures present in eukaryotic cells, such as lung epithelial cells and palisade cells, that are not present in prokaryotic cells. [3]

> Measure the length in millimetres, and divide by the magnification. Convert to micrometres by multiplying by 1000. Write down the steps in your calculation.

(e) Explain how the cells lining the lung are adapted to allow efficient exchange of gases. [3]

2 The table shows chemical links formed between the glucose units of some carbohydrates.

> In Q1(b) make sure that you write down three points about the **structure** of a palisade cell.

carbohydrate	chemical links between the glucose units		
	1–4 α glycosidic	**1–4 β glycosidic**	**1–6 α glycosidic**
maltose	✓	✗	✗

(a) Draw a table similar to the one above to show whether or not the type of links are found in the following carbohydrates: amylose, amylopectin, cellulose and glycogen. Maltose has already been completed for you. [4]

> In Q2(b) think about the steps that you would follow and the variables that must be controlled to make a valid comparison.

(b) Explain how you would show that a sample of grape juice contains more reducing sugar than a sample of apple juice. [4]

3 (a) State the elements found in **all** amino acids. [2]

(b) Name the type of chemical bond that links the amino acids; [1]

(c) Name one fibrous protein and one globular protein. [2]

(d) Describe how a polypeptide chain forms a globular structure. [4]

(e) Describe how the structure of collagen (a fibrous protein) differs from the structure of cellulose (a carbohydrate). [7]

> Q3(e) and Q4(b) require an extended answer and there will be one mark for the way in which you write your answer using specialist terms in the correct way.

4 (a) Explain how an enzyme catalyses a reaction. [4]

   (b) Describe and explain the effects of temperature on enzyme activity. [7]

Question 4(b): an easy way to describe the effects is to sketch a graph similar to the one on page 22. This may help you to write a description.

5 Nucleic acids (DNA and RNA) are large molecules.

   (a) State the name of the small molecules that join together to form nucleic acids. [1]

   (b) State **three** structural differences between DNA and RNA. [3]

In Q5(b) you need to say DNA is....., but RNA is.......

DNA was extracted from pig liver cells and from yeast cells. The DNA from the two sources was broken down and the quantities of the four bases, adenine, guanine, thymine and cytosine, were measured. The table below shows the percentage composition of the four bases found in the DNA extracted from pig liver cells and from yeast.

source of DNA	adenine(%)	guanine(%)	thymine(%)	cytosine(%)
pig liver cells	29.4	20.5	29.7	20.5
yeast	31.3	18.7	32.9	17.1

   (c) Calculate the ratios of (i) adenine to thymine, and (ii) guanine to cytosine in the DNA from pig liver cells and from yeast. [4]

   (d) Use the data and the ratios you have calculated to explain the importance of base pairing in the structure and replication of DNA. [7]

Q5(c) says 'use the data', so there are marks for putting some of the data in your answer to illustrate the points you make.

6 (a) Name **two** proteins, used to treat disease, that are produced commercially by genetically modified cells. [2]

   (b) Describe the stages that are involved in producing bacteria that can make human proteins. [8]

7 (a) State the functions during mitosis in animal cells of (i) centrioles, and (ii) centromeres. [3]

   (b) Replication of DNA occurs during interphase in the cell cycle. If a cell starts the process of mitosis before DNA replication is complete, both daughter cells die. Explain why this is so. [2]

Q6(b) requires an extended answer and there will be one mark for the way in which you write your answer using specialist terms in the correct way.

The sequence of events in the life cycles of some animals is very different from those in humans. For example, all the cells in the body of an adult male ant are haploid.

   (c) Explain what is meant by the term *haploid*. [2]

   (d) State and explain whether sperm production in ants involves mitosis or meiosis. [2]

   (e) State how many genetically different types of sperm cell can be produced by a male ant. Explain your answer. [2]

Do not be surprised to find some information on something unfamiliar such as the sex life of ants!

8 (a) Explain why large carnivores, such as lions, tigers and birds of prey, are not as common as the animals on which they feed. [4]

   (b) Many species of bacteria are involved in cycling nitrogen in terrestrial ecosystems. Describe the roles of the bacteria *Rhizobium*, *Nitrosomonas* and *Nitrobacter* in the nitrogen cycle. [8]

Q8(b) requires an extended answer and there will be one mark for the way in which you write your answer using specialist terms in the correct way.

Module 2803/01: Transport

Transport systems in animals

Very small organisms like *Stentor* (see page 6) do not have a specialised transport system. Oxygen from its surroundings diffuses through the cell surface membrane and the cytoplasm to mitochondria where it is used in respiration. The distance from the edge of the cell to the centre is no more than 0.5 mm. Large organisms cannot rely on diffusion from the surface for their supplies of oxygen. There are two reasons for this: the body surface is not large enough; distances from the surface to the centre are too great.

Surface area to volume ratios

Imagine that an organism has the shape of a cube. Also imagine that its body grows from a small cube that has sides of 1 mm into a larger one with sides of 10 mm. Study the table carefully and you will see that as the organism grows its volume (or mass) increases faster than its surface area. A large organism has less surface area for each mm^3 of body than a small one. This means that there is less space on the body surface for the uptake of oxygen and the removal of carbon dioxide by diffusion. Many organisms therefore have special surfaces for gas exchange such as gills and lungs.

	length of side (mm)	volume (mm^3)	surface area (mm^2)	surface area: volume ratio
	1	1	6	6:1
	5	125	150	1.2:1
	10	1000	600	0.6:1

◖ Few cells are really this large. We are using 'cells' of this size to show you the idea of the ratio.

Many animals have transport systems to move oxygen from gas exchange surfaces to cells deep in the body. Diffusion is only effective over short distances, such as a few micrometres. In mammals, oxygen diffuses across the surface of the alveoli in the lungs and enters the blood which is pumped to the rest of the body by the heart. As blood flows around the body nutrients, oxygen, carbon dioxide and other waste compounds are exchanged with the tissues.

Transport in a mammal

The transport system of mammals consists of:
- blood – a suspension of cells;
- vessels – arteries, veins and capillaries;
- a pump – the heart.

It is a **mass flow system** because all the blood flows in the same direction through a system of vessels. It is a **closed blood system** because the blood flows inside vessels. Insects have an open system since their blood is not enclosed in vessels and circulates in body spaces. Mammals have a **double circulation** because blood flows through two circuits:
- pulmonary circuit – from the heart to the lungs and back;
- systemic circuit – from the heart to the rest of the body and back.

When the heart pumps blood it gives it a pressure that forces it through the blood vessels. It is difficult to push blood through the vessels so the heart gives the blood considerable pressure (see page 48).

Blood vessels

Substances, such as oxygen, carbon dioxide and glucose, are exchanged between blood and tissues through the walls of capillaries. Blood flows from the heart through arteries to reach capillaries and then returns to the heart inside veins. All blood vessels are lined by squamous epithelium.

blood vessel (not to scale)	how structure is related to function
artery outer layer of collagen fibres middle layer of muscle and elastic fibres inner layer of elastic fibres and squamous cells lumen actual diameter = 2 mm	• walls are thick and strong to withstand high blood pressure • elastic fibres stretch when heart pumps blood into an artery • elastic fibres recoil to push blood on its way towards capillaries • muscle in smaller arteries controls diameter to alter blood flow
capillary squamous cells forming lining of capillary nucleus red blood cell lumen actual diameter = 8 μm	• wall is made of one layer of squamous cells so diffusion distance is short • tiny holes in endothelial cells allow water and some solutes to leave the blood • very small, so many capillaries in a small space gives a large surface area for diffusion
vein thin layer of collagen fibres thin layer of muscle and elastic fibres lumen actual diameter = 10 mm	• walls are thin as blood pressure is low • walls distend (stretch) to accommodate large volumes of blood • veins have semilunar valves to ensure blood travels towards the heart
towards heart valve open valve closed	• pressure of blood forces valves open • backflow of blood closes the valve • muscles of the body contract, e.g. during walking, and squeeze veins so pushing blood towards the heart

✓ *Quick check 1, 2, 3*

❓ Quick check questions

1 Calculate the magnification of the cross sections of the artery, capillary and vein in the diagram above.

2 Make a table to compare the structure and functions of the three types of blood vessel. Use these column headings:

feature	artery	capillary	vein

3 Explain how the structures of arteries, capillaries and veins are related to their functions.

Blood, tissue fluid and lymph

- Blood is a suspension of red and white cells and platelets in plasma. When left to settle or spun in a centrifuge, blood separates into these three components.

- Tissue fluid is a colourless fluid that is formed from blood plasma by pressure filtration through capillary walls. It surrounds all the cells of the body and all exchanges between blood and cells occur through it.

- Lymph is tissue fluid that has drained into lymphatic vessels. It passes through lymph nodes where it gains white cells and antibodies. Lymphatic vessels absorb hormones from some endocrine glands and fat in the small intestine.

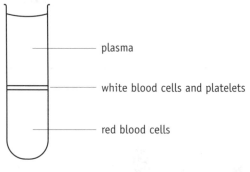

plasma

white blood cells and platelets

red blood cells

Fig. 21.1 The composition of blood.

The table below shows the structure of the blood cells as seen through the light microscope.

cell	relationship between structure and function
red blood cell 7 μm	• biconcave disc shape gives a large surface area for diffusion of oxygen and carbon dioxide; • no organelles, so cytoplasm is full of haemoglobin; • elastic membrane allowing cells to change shape as they squeeze through capillaries and restore shape when enter veins.
phagocyte (neutrophil) 9 μm	• phagocytosis – bacteria engulfed in vacuoles and digested; • large number of lysosomes for digestion of bacteria; • lobed nucleus to help squeeze through gaps between cells in capillary walls.
lymphocyte 5.5 μm	• some lymphocytes develop into plasma cells that have large quantity of rough endoplasmic reticulum for fast production of antibodies.

✓ *Quick check 1, 2*

This table summarises the differences between blood, tissue fluid and lymph

component	blood	tissue fluid	lymph
red blood cells	✓	✗	✗
white blood cells	✓	some	some
water	✓	✓	✓
plasma proteins	✓	very few	very few
sodium ions	✓	✓	✓
glucose	✓	✓	very little
antibodies	✓	✓	✓
fats	✓	some	✓ especially after a meal

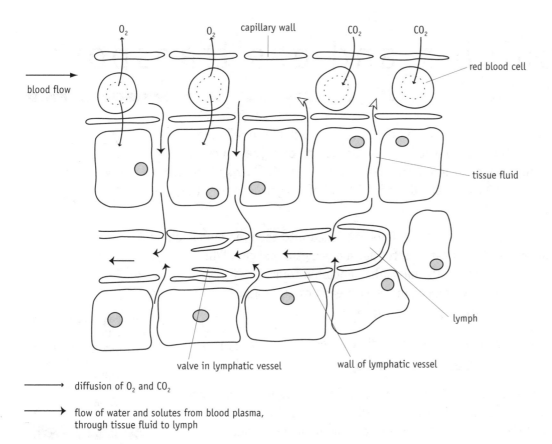

→ diffusion of O_2 and CO_2

→ flow of water and solutes from blood plasma, through tissue fluid to lymph

⇀ flow of water into blood plasma by osmosis

Fig. 21.3 Tissue fluid bathes all the cells of the body. It is formed from blood plasma and drains into lymphatic vessels to form lymph.

Blood enters capillaries at a relatively high blood pressure. This forces small molecules such as water, ions and glucose, through the small holes in the walls of the endothelial cells to form tissue fluid. Some of the plasma proteins also leave the blood. Much of the water returns to the blood by osmosis, but not all. Excess fluid returns to the blood through the lymph vessels and this ensures that tissues do not fill with too much fluid. When that does happen a person suffers from oedema which can be very dangerous if it happens in organs such as the lungs.

✓ *Quick check 3*

? *Quick check questions*

1 Calculate the magnification of the red blood cell shown on page 44.

2 Explain how the structure of red blood cells is related to their function.

3 Describe how tissue fluid differs in composition from blood. Explain how tissue fluid is formed.

Haemoglobin and gas transport

Oxygen is not very soluble in water. If we did not have haemoglobin in red blood cells to transport oxygen, the blood would carry about 0.3 cm^3 of oxygen in every 100 cm^3. This is how much dissolves in water. When blood leaves the lungs it carries far more than this – every 100 cm^3 of blood carries 20 cm^3 of oxygen. Almost all of this is combined with haemoglobin.

There are over 280 million molecules of haemoglobin packed into each red blood cell and each haemoglobin molecule can carry up to four molecules of oxygen. When blood flows through the capillaries in the lungs, haemoglobin forms oxyhaemoglobin. A molecule of oxygen combines with each haem group:

$$Hb + 4O_2 \longrightarrow HbO_8$$

In the tissues this reverses and oxyhaemoglobin dissociates to give up oxygen:

$$HbO_8 \longrightarrow Hb + 4O_2$$

Haemoglobin also transports carbon dioxide. As blood flows through tissues, some carbon dioxide molecules react with free amino (–NH_2) groups at the ends of the α and β polypeptides to form a compound known as carbaminohaemoglobin.

> ◖ Haemoglobin is a globular protein and has four haem groups for binding oxygen – see page 14.

> ◖ Do not confuse this with *carboxyhaemoglobin* which forms when carbon *monoxide* combines with haemoglobin.

> ✓ *Quick check 1*

Dissociation curves

Dissociation curves show how efficient haemoglobin is at absorbing oxygen in the lungs and delivering oxygen to tissues. Samples of blood are exposed to different mixtures of oxygen and nitrogen and shaken to ensure that haemoglobin absorbs as much oxygen as possible. Partial pressure (measured in kilopascals or kPa) is the pressure exerted by oxygen. The % saturation is calculated as the % of the *maximum* quantity of oxygen that haemoglobin absorbs.

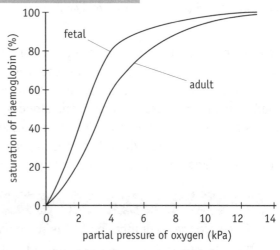

Fig. 22.1

> ◖ You can use haemoglobin dissociation curves to predict the saturation of the blood with oxygen in different parts of the circulation.

	partial pressure of oxygen (kPa)	saturation of haemoglobin with oxygen (%)
lungs (at sea level)	13.0	98
respiring tissues	5.0	70
actively respiring tissues (e.g. during exercise)	3.5	50
maternal blood in the placenta	4.0	60
fetal blood in the placenta	4.0	80
lungs at an altitude of 6500 metres	5.0	70

Put a ruler along the vertical axis on Fig. 21.1. Move the ruler along the graph from 0 kPa to 14 kPa. An important feature is the sigmoid or S-shape of the curve. Position the ruler on the graph to confirm the readings in the table. Confirm the observations that follow.

- Haemoglobin is fully saturated at the partial pressure of oxygen in the lungs. It becomes fully loaded with oxygen as it passes the gas exchange surface (the alveoli).

- As it flows through the tissues, haemoglobin responds to low partial pressures of oxygen by dissociating (giving up some of its oxygen). Oxygen diffuses through capillary walls and tissue fluid to respiring cells.

- As tissues use up oxygen during exercise, haemoglobin dissociates even more.

- The S-shape of the curve shows that haemoglobin responds to small *decreases* in oxygen concentration in the tissues by giving up a *lot* of oxygen.

- Fetal blood has a higher affinity for oxygen than adult blood. This means that oxygen diffuses from maternal blood to fetal blood across the placenta, even though the partial pressures are similar.

- Haemoglobin is only 70% saturated when we travel to places at altitudes of 6500 metres.

✓ *Quick check 2*

When cells respire they produce carbon dioxide. Fig. 22.2 shows what happens when carbon dioxide is added to the gas mixture. Take a ruler and put it at the point marked Z on the graph. When there is more carbon dioxide in the mixture, haemoglobin is less saturated with oxygen.

Now put the ruler parallel with the horizontal axis and see that curve B is to the right of A. This effect of carbon dioxide on the dissociation curve is the **Bohr effect**.

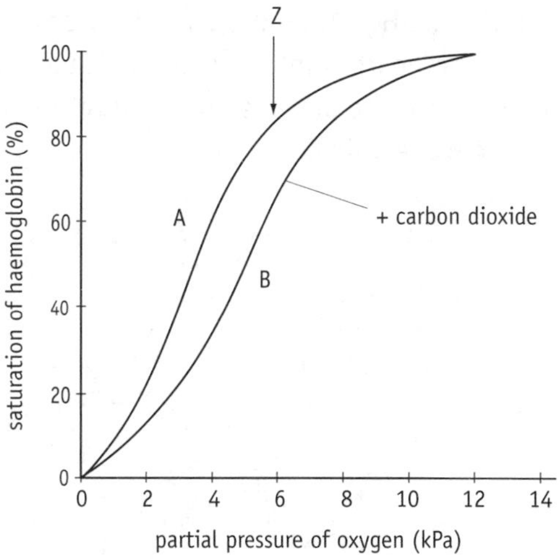

Carbon dioxide interacts with haemoglobin to cause it to give up its oxygen. This is good news, because it means that haemoglobin delivers more oxygen to those tissues that are respiring fast, such as muscles during exercise. This happens because an enzyme in red blood cells, carbonic anhydrase, catalyses the reaction between water and carbon dioxide:

Fig. 22.2

$$H_2O + CO_2 \xrightleftharpoons{\text{carbonic anhydrase}} H_2CO_3 \rightleftharpoons H^+ + HCO_3^-$$

Haemoglobin absorbs the hydrogen ions that form inside the red blood cells and this causes them to lose the oxygen molecules that they are carrying.

✓ *Quick check 3*

Gasping for breath

The air pressure is much lower at high altitude than at sea level and although 20% of the air is still oxygen, the partial pressure is much less. This causes shortness of breath. To compensate for this more red blood cells are made and the percentage increases from 45% of blood volume to as much as 70%. This acclimatisation takes about a week.

✓ *Quick check 4*

? Quick check questions

1 Describe how haemoglobin transports oxygen and carbon dioxide.

2 Explain the advantage of the S-shaped dissociation curve for haemoglobin.

3 Describe and explain the Bohr effect.

4 Explain why the number of red blood cells increases in people who travel to places at high altitude.

Heart – structure and function

Put a stethoscope to someone's chest and you will hear a familiar noise – sometimes described as 'lub-dup'. This is the noise of the valves closing in the heart during a heart beat. One beat of the heart pumps blood through the pulmonary and systemic circuits. The heart has two pumps working in series. The right side of the heart pumps deoxygenated blood to the lungs through the pulmonary artery at a blood pressure of about 24 mmHg (3.2 kPa). The left side pumps oxygenated blood into the aorta at about 120 mmHg (15.8 kPa). The flow of blood through the heart is intermittent as it pushes blood out into the arteries and then refills with blood from the veins.

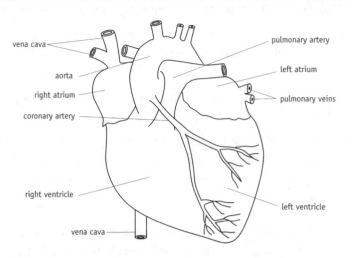

Fig. 23.1 This is the external view of the human heart showing the four chambers, the main blood vessels and the coronary arteries that supply blood to heart muscle.

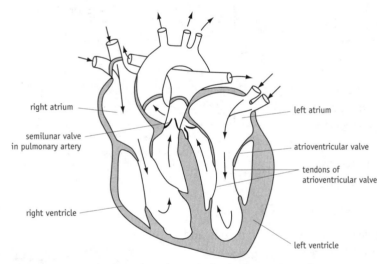

Fig. 23.2 A vertical section through the heart to show the internal structure. The left and right atria have much thinner walls than the ventricles because the atria contract to push blood a short distance at a low pressure into the ventricles.

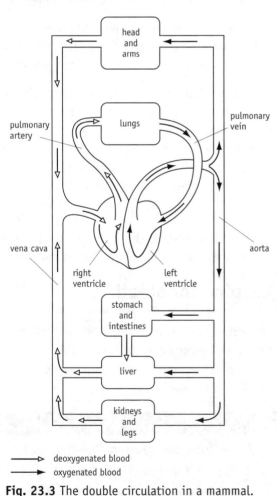

Fig. 23.3 The double circulation in a mammal.

The wall of the left ventricle is much thicker than that of the right ventricle because it contracts to force blood into the aorta at high pressure as the blood in the systemic circuit meets much more resistance to flow. The lungs receive blood from the right ventricle. The lungs are very spongy and the blood vessels allow blood to flow easily for maximum exchange of gases in the alveoli.

✔ *Quick check 1*

Initiating the heart beat

The heart is made of cardiac muscle which is myogenic, which means that it contracts of its own accord. The sinoatrial node (or SAN), which is in the wall of the right atrium, is the heart's natural pacemaker. It sends out a stream of electrical impulses to initiate each heart beat. These impulses spread across the atrial muscle which contracts first. Impulses reach the atrioventricular node (AVN) which is between the atria and ventricles. This is like a relay station that delays the impulses for a short while so the ventricles do not contract too soon. The AVN sends the impulses along special, fast conducting muscle cells in the septum called Purkyne fibres to the base of the heart. Muscle at the base of the ventricles contracts and pushes the blood upwards into the arteries. Nerves to the heart change how fast or slow the heart beats.

✓ *Quick check 2*

The cardiac cycle

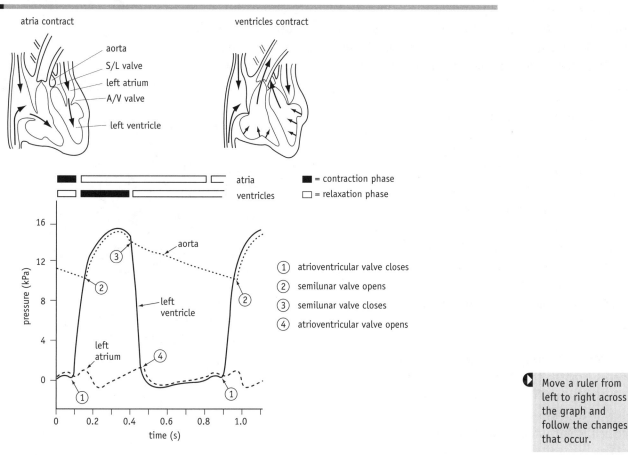

Fig. 23.4 Cardiac cycle

Fig. 23.4 shows the changes in blood pressure in the left atrium, left ventricle and aorta during one heartbeat and the beginning of the next.

✓ *Quick check 3,4,5*

❓ Quick check questions

> Move a ruler from left to right across the graph and follow the changes that occur.

1 Describe the pathway taken by blood as it flows through the heart.

2 Explain why the ventricles contract from the base.

3 How long is one heart beat? What is the heart rate in beats per minute?

4 Explain why the atrioventricular valve closes at 1 and opens at 4.

5 Explain why the semilunar valve opens at 2 and closes at 3.

> Look at Fig. 23.4 before answering Quick check questions 3, 4 and 5

Transport in plants

Plants use photosynthesis to convert light energy to chemical energy in compounds such as carbohydrates, fats and proteins and many more. Simple inorganic substances, such as carbon dioxide, water and ions from their immediate environment are used as raw materials to make these assimilates.

Plants have two separate transport tissues

Water and ions travel upwards in **xylem** tissue from the roots to stems, leaves, flowers and fruits. Sucrose and other assimilates travel upwards and downwards in **phloem** tissue. On page 7, you saw where these tissues are located inside plant organs. Movement of water in xylem and assimilates in phloem is by **mass flow**. Everything travels in the same direction within each column of xylem or phloem cells.

✓ *Quick check 1*

Plants do not have a transport tissue for oxygen and carbon dioxide. These gases diffuse through air spaces to and from cells. This is possible because diffusion of oxygen, for example, through air is 10 000 times faster than diffusion through water or through cells. There are air spaces between cells in all plant organs which allow the gases to reach even the centre of a thick root. As plants have a much lower metabolic rate and a large surface area to volume ratio than animals, diffusion is sufficient. Unlike animals, plants do not have an equivalent of haemoglobin for oxygen transport. The cell surfaces exposed to air spaces form a large gas exchange surface. You can see this clearly in a cross-section of a leaf where each cell is in contact with the air (Fig. 24.1).

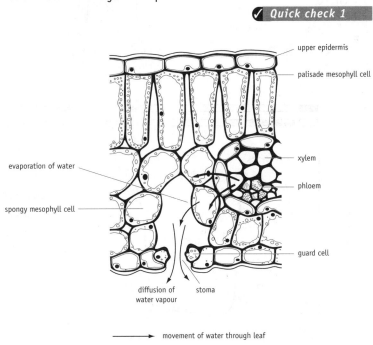

Fig. 24.1

Transpiration is inevitable

Plants lose huge amounts of **water vapour** to the atmosphere. This loss of water vapour is transpiration:

- water **evaporates** from moist cell walls;
- water vapour **diffuses** from the air spaces inside leaves to the atmosphere.

Palisade and spongy mesophyll cells have a very large internal surface for gas exchange. Since the carbon dioxide concentration in the air is so low (0.04% or 400 parts per million), the surfaces are large so that enough can be absorbed for photosynthesis.

✓ *Quick check 2*

The air inside leaves is always fully saturated with water vapour. Usually, the air outside is less saturated than this and so a concentration gradient for water vapour exists between the air spaces and the outside. Water vapour therefore diffuses down this humidity gradient. The pathway with the least resistance is through the stomata. As stomata are open during the day to allow carbon dioxide to diffuse in, water vapour diffuses out. At night, most plants close their stomata, so the rate of water loss decreases.

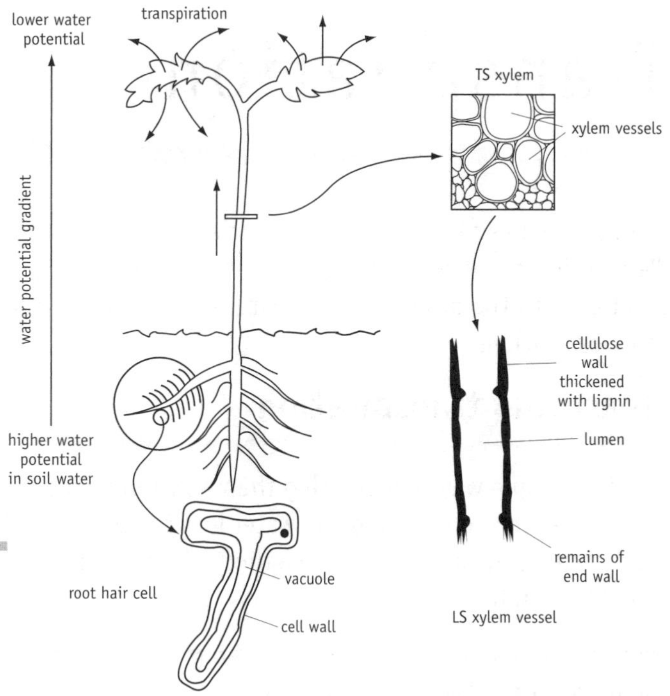

Transpiration drives the movement of water in plants

The loss of water from leaves by transpiration causes water to travel upwards through the plant by mass flow. The mechanism is called 'cohesion-tension' and it works as follows:

Fig. 24.2

- Water molecules are 'sticky' as they are attached to each other by hydrogen bonds. This mutual attachment is called **cohesion**.
- Water evaporates from mesophyll cells and this lowers their water potential.
- Water moves along gradients of water potential, from the high water potential in the xylem to the lower potential in the mesophyll cells.
- Continuous columns of water molecules are pulled from the xylem towards the mesophyll as a result of forces of cohesion.
- The pull on columns of water molecules extends all the way to the roots and into the soil.
- Water molecules are also attracted by hydrogen bonds to cellulose in cell walls. This is **adhesion**. This helps maintain columns of water in the xylem when there is little transpiration.
- The energy for water movement comes from the thermal energy evaporating water from the mesophyll surface.

Xylem vessels are adapted for transport of water

Xylem vessels are hollow, dead cells that have no end walls and no cytoplasm. They form a 'plumbing system' that offers little resistance to the flow of water. A tough substance called lignin strengthens the cellulose cell walls. This prevents xylem vessels collapsing inwards due to the tension that develops, especially on hot days when rates of transpiration are high.

> **▶** Note that root hairs have:
> - a large surface area;
> - thin cell walls;
> - lower water potential than soil water;
> - carrier molecules in cell membranes for absorption of ions.

> **▶** Note the importance of hydrogen bonds again – see page 16 if you are not sure what they are.

✓ *Quick check 3*

✓ *Quick check 4*

Note the importance of hydrogen bonds again – see page 16 if you are not sure what they are.

? *Quick check questions*

1 Explain why movement of water in plants is an example of mass flow.
2 Define the term **transpiration**.
3 Describe the pathway taken by water as it moves through a plant from the soil to the atmosphere.
4 Explain how xylem vessels are adapted to the transport of water.

Transpiration

Most plants need considerable quantities of water to keep them alive
We can find out how much water plants take up through their roots
and lose into the atmosphere as water vapour by carrying out a varie
of investigations.

Measuring transpiration

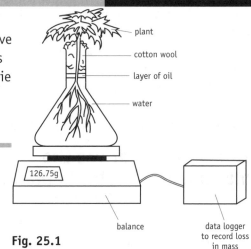

Fig. 25.1

This is a simple way of measuring the water uptake by marking the
water level on the flask and the water loss by following the loss in
mass. The disadvantage of this method is that it takes a long time
to obtain results.

Many school laboratories have potometers like this to measure
the uptake of water by leafy shoots.

1 Cut a shoot under water.

2 Place the shoot in the potometer under
    water. *This avoids getting any air into the
    stem, which would block the xylem vessels
    so that water does not flow.*

3 Allow time for the plant to adjust to the
    surroundings.

4 Open the tap on the reservoir to position
    the air bubble at the end of the scale.

5 Keep the following environmental
    conditions constant around the plant:
    light intensity, humidity, temperature,
    air movement.

6 Take readings by timing how long it takes the air
    bubble to move a set distance along the scale.

7 Reset the air bubble with water from the reservoir
    and take more readings.

8 Take the mean of at least three results.

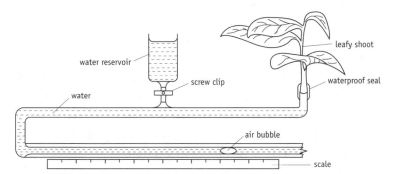

Fig. 25.2 Potometer

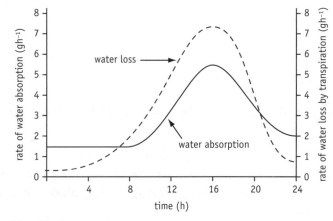

Fig. 25.3

As most of the water absorbed by the plant is lost
to the atmosphere it is possible to assume that this
is also the rate of transpiration. However, water is
also used by the plant as a raw material in
photosynthesis and for keeping cells turgid. If the
potometer is placed on a sensitive balance that
records very small changes in mass, then the rate of
transpiration can be measured at the same time. Fig 25.3 shows how much water is
absorbed by a plant in a potometer and lost by transpiration to the atmosphere.

✓ *Quick check 1*

Note that:

1 rates of water absorption and water loss are highest around midday.

2 rates of transpiration are low at night because stomata are closed.

3 water continues to be absorbed at night because so much water has been lost
    during the day that more is needed to maintain turgidity of cells.

Environmental factors influence rates of transpiration

Experiments with potometers show that light intensity, humidity, temperature and air movement influence rates of transpiration.

Stomata are usually closed when it is dark. As the **light intensity** increases, stomata open wider, so more water vapour can escape. After a certain light intensity, the stomata cannot open any wider so the rate of transpiration remains constant.

The **humidity** of the air surrounding a plant determines the steepness of the concentration gradient between the air spaces in the leaf and the atmosphere. If the air outside is dry, then the maximum gradient exists. If the air is fully saturated (like the air inside the leaf) there is no gradient and there is no net loss of water vapour through the stomata.

Temperature determines rates of evaporation inside leaves. The water holding capacity of the atmosphere outside the leaf is also determined by temperature. High rates of transpiration therefore occur on hot days.

When there is no **air movement** around a leaf, water vapour molecules collect around the leaf surface and the air is saturated. Little or no transpiration occurs. When air blows over the surface of a leaf it carries water vapour molecules away so rates of transpiration are high.

Xerophytes

These are plants adapted to survive in dry places, such as deserts. They have a range of adaptations to reduce the loss of water vapour by transpiration:

- small leaves to reduce the surface area;
- thick leaves to reduce surface area : volume ratio;
- stomata set deep inside the leaf so that they are at the base of a depression full of water vapour;
- thick waxy cuticles to reduce water loss through the epidermis.

Some desert plants open their stomata at night, absorb and store carbon dioxide and then close their stomata during the day. This reduces water loss considerably.

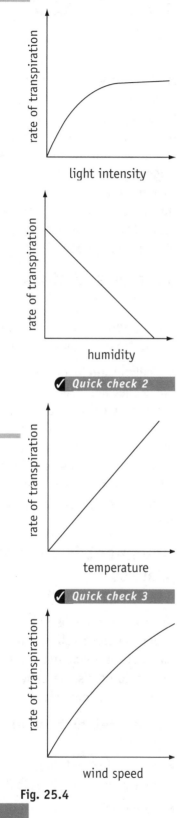

✓ Quick check 2

✓ Quick check 3

Fig. 25.4

✓ Quick check 2

✓ Quick check 3

> ### ❓ Quick check questions
>
> 1 Outline the steps you should take when using a potometer to measure rates of transpiration at different air speeds. Explain any precautions that are necessary.
>
> 2 State four environmental factors that influence rates of transpiration.
>
> 3 Define the term **xerophyte**.

Transport in the phloem

Most photosynthesis in plants occurs in leaves, which use carbon dioxide and water as raw materials for making three carbon sugars known as trioses. These substances are used as the basis for many different compounds. All the reactions of photosynthesis occur in chloroplasts. These organelles convert simple sugars into amino acids, using nitrate or ammonium ions (NH_4^+), and make chlorophyll using magnesium ions. Leaf cells also convert trioses into glucose and fructose and combine these two to make sucrose. Sucrose is the main molecule used by plants for transport. Compounds that plants have made from simple raw materials are called **assimilates**. Many of these assimilates are exported from leaves to the rest of the plant in the **phloem**.

Sources and sinks

The transport of assimilates in phloem tissue is called **translocation**, which literally means 'from one place to another'.

Assimilates are loaded in the phloem in the leaves where they are made. Leaves are the source of the sugars and amino acids and are often called **sources**. Assimilates are transported to other parts of the plant, which use the assimilates to provide energy or materials for synthesising macromolecules such as cellulose and proteins. These places, such as roots, stems, flowers, fruits and seeds, are referred to as **sinks.**

Movement in the phloem is an active process

Sucrose and other assimilates travel throughout a plant in phloem **sieve tubes**, which are made from cells called **sieve elements**. Alongside sieve tubes are **companion cells**. Mesophyll cells in the leaf are close to veins containing sieve tubes.

Sucrose travels to the phloem companion cells in two ways:

- from cell to cell through narrow tubes of cytoplasm that penetrate cell walls (known as plasmodesmata);
- along cell walls in the mesophyll.

Carrier proteins in the cell surface membranes of companion cells actively pump sucrose into the cytoplasm. From here it passes through plasmodesmata into a sieve element.

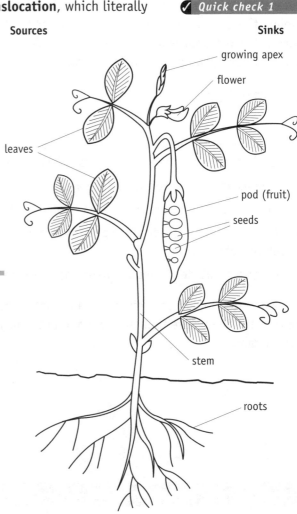

✓ *Quick check 1*

Fig. 26.1 Sinks and sources in a pea plant.

The accumulation of sucrose and other solutes, such as amino acids, in sieve elements lowers the water potential so that water diffuses in by osmosis from adjacent cells and from the xylem. This creates pressure in the sieve elements causing the liquid (phloem sap) to flow out of the leaf.

Phloem sieve elements are adapted for transport as they have:

- end walls that have sieve pores allowing phloem sap to flow freely;
- little cytoplasm to impede the flow of sap;
- plasmodesmata to allow assimilates to flow in from companion cells.

Sieve elements differ from xylem vessels because they are alive: they have some cytoplasm with organelles. They are not lignified, as they do not need to withstand the same forces of tension that exist in xylem.

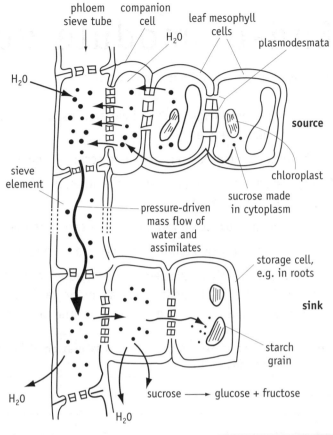

Fig. 26.2

✓ *Quick check 2*

Sucrose is unloaded at sinks. It is not known how this is done, but it seems likely that an enzyme breaks down sucrose to glucose and fructose, which are taken up by cells and respired or stored as starch. This reduces the solute concentration of phloem sap and lowers the pressure, so helping to maintain a pressure gradient from source to sink so the sap keeps flowing in the phloem.

If sections of phloem are cooled or a substance that inhibits respiration is added to phloem, transport slows or stops completely. This suggests that transport in phloem is an active process.

✓ *Quick check 3*

? *Quick check questions*

1 Define the following terms: **assimilate** and **translocation**.
2 Describe how phloem sieve tube elements are adapted for the transport of sucrose.
3 Explain how sucrose travels from a source to a sink.

End-of-module questions

1 (a) Explain why large, multicellular animals, such as mammals, need a blood circulatory system. [3]

   (b) Describe how gaseous exchange occurs in the alveolus. [4]

2 (a) Describe *three* ways in which the structure of an artery differs from the structure of a vein. [3]

   (b) Explain how blood returns to the heart in the veins. [4]

   (c) Explain how tissue fluid is formed from blood plasma. [3]

3 The table below shows the percentage saturation of haemoglobin with oxygen when it is exposed to tissue fluids which vary in the partial pressure of dissolved oxygen. The tissue fluids are at two different concentrations of carbon dioxide (**A** and **B**).

partial pressure of oxygen in tissue fluids (kPa)	% saturation of haemoglobin with oxygen	
	A	**B**
0	0	0
2	14	26
4	39	54
6	66	75
8	80	85
10	88	92

> You will not be expected to draw graphs in an AS theory exam paper. Drawing this graph may help you to understand this topic.

(a) Draw a graph of the data given in the table.

(b) With reference to your graph,

> 'With reference to...' means that you need to study the graph carefully and give figures in your answers.

   (i) state what happens to the percentage saturation of haemoglobin as the partial pressure of oxygen in tissue fluids increases in **B**; [2]

   (ii) state which curve (**A** or **B**) shows tissue fluid with the higher concentration of carbon dioxide; [1]

   (iii) explain why the carbon dioxide concentration may increase in the tissue fluid in muscles; [2]

   (iv) state the name used to describe the effect of carbon dioxide on the percentage saturation of haemoglobin with oxygen and explain the advantage of this effect during strenuous exercise. [4]

Fetal haemoglobin has a greater affinity for oxygen and is fully saturated at lower partial pressures of oxygen than adult haemoglobin.

   (c) Explain the significance of this. [3]

4 The mammalian heart is described as a double pump. Heart muscle contracts of its own accord, it does not rely on nerves to initiate a heart beat.

   (a) Explain the advantages of having a double pump. [3]

   Fig. 27.1 shows the position in the heart of the sinoatrial node, the atrioventricular node and Purkyne tissue.

(b) Describe the roles of the SAN, AVN and Purkyne tissue in initiating a heart beat. [3]

(c) Describe the events that occur during the cardiac cycle highlighting the action of the valves. [8]

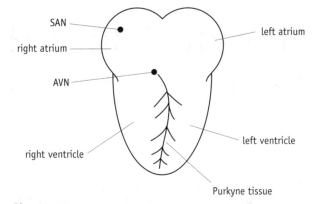

Fig. 27.1

5 Transpiration is the loss of water vapour by leaves.

(a) Explain why water loss from leaves is inevitable. [4]

(b) Describe the pathway taken by water molecules as they move from the xylem *in the leaf* to the air outside the leaf. [4]

Q4(c) requires an extended answer and there will be one mark for the way in which you write your answer.

(c) Explain, *in terms of water potential*, the movement of water that you have described. [4]

An experiment was carried out on a tropical vine which grows by climbing around trees. The stem was cut and placed into a bucket of water coloured with a red dye. After a while the veins in the leaves, 50 m above the ground, were stained red.

(d) Explain how water with the coloured dye moved up the stem of the vine to reach the leaves. [4]

6 Transport in plants and animals often uses a process known as mass flow.

(a) Explain the meaning of the term **mass flow**. [1]

(b) With reference to transport of assimilates, such as sucrose, in plants, explain the difference between a **source** and a **sink**. [4]

(c) Explain how phloem sieve elements are adapted for the transport of assimilates. [4]

7 Fig. 27.2 shows an investigation into the transport of substances in a strawberry runner. The mature enclosed leaf was brightly illuminated. All other parts of the plant were kept in darkness. After ten minutes exposure to $^{14}CO_2$, the flask was removed. The leaf which had been enclosed was found to be radioactive and after a further ten minutes, the young plant became radioactive.

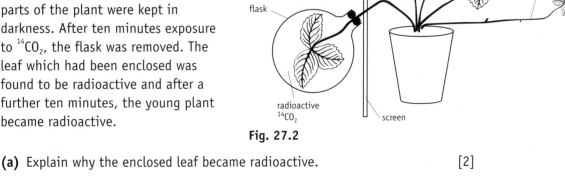

Fig. 27.2

(a) Explain why the enclosed leaf became radioactive. [2]

(b) Explain how radioactivity reached the young plant. [3]

Module 2802: Human health and disease

Health and disease

The World Health Organisation defines **health** as a **state of complete physical, mental and social well-being which is more than just the absence of disease**.

Disease may be defined as an absence of health, but we tend to use the word to refer to specific states of bad health that give certain symptoms that we experience. We report these symptoms to doctors who look for certain clinical signs to decide which disease we have.

You may have studied some diseases as part of your GCSE course and you will be familiar with others. There are several different types of diseases and different ways of classifying them into groups. Most diseases fit into several of these groups or categories. Work through this flow chart to classify the diseases listed in this module.

> ◖ Try using the flow chart with other diseases that you know about.

In addition to the seven categories of disease shown in the flow chart there are two more:

- **Social diseases:** these are influenced by social conditions and people's way of life and environment. Many diseases fall into this category, even infectious ones. The increase in the incidence of tuberculosis (TB) in recent years is partly due to an increase in overcrowding in poor housing and urban decay in cities, such as New York and London.

- **Self-inflicted diseases:** these are caused by the choices people make about their way of life. Some people misuse drugs such as alcohol and nicotine which may contribute to diseases, such as cirrhosis of the liver, lung cancer and coronary heart disease. More obvious examples are attempts at suicide and starving oneself.

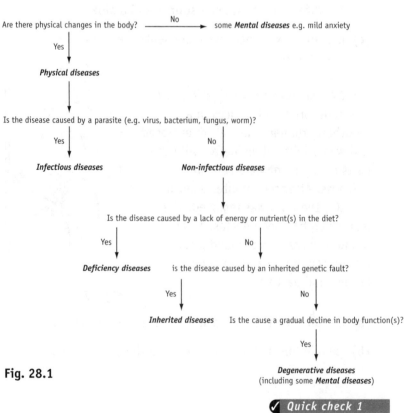

Fig. 28.1

✓ *Quick check 1*

Epidemiology

This is the study of patterns of disease. Epidemiologists collect health statistics for populations, such as those living in the UK. The data is presented for certain time periods, e.g. per year.

- **Incidence** – the number of people who are diagnosed with a disease.
- **Death rates** – the number of people who die from a certain disease.
- **Prevalence** – the number of people who have a certain disease.

Diseases that are always present in a population are described as **endemic.** When the incidence of a disease increases so that large numbers of people suffer from its effects, there is an **epidemic.** When a disease spreads across a whole continent, or across the world, there is a **pandemic.**

✓ *Quick check 2*

Collecting health statistics is useful for:
- finding out which diseases are most common;
- deciding on priorities for health care;
- comparing populations living in different parts of a country;
- comparing populations in different countries.

These statistics tell us that people in developing countries are at risk of infectious diseases, such as malaria, TB and cholera, while people in developed countries are most at risk of degenerative diseases, such as coronary heart disease (CHD), strokes and cancers. As the economic status of a country improves, people are more likely to die from degenerative diseases than infectious diseases.

The Human Genome Project

This huge worldwide project involves identifying all human genes, finding their positions on chromosomes and sequencing the bases (A,C,T and G) in human DNA. Some of the advantages for health are the following.

- Genetic tests can be developed for inherited diseases.
- The likelihood that people will develop disease can be assessed (susceptibility to disease is strongly influenced by genes even when not caused by a genetic defect).
- Doctors will be able to make a more accurate diagnosis and choose more appropriate treatments avoiding drugs likely to have side effects.
- Pharmaceutical companies may be able to develop drugs better able to target specific problems.
- Medical researchers will have more data when looking for causes of disease and finding cures.

Look at page 33 to see the importance of the base sequences. They determine the sequence of amino acids in proteins.

✓ *Quick check 3*

❓ Quick check questions

1 Use the flow chart to classify the following diseases: measles, cholera, coronary heart disease, rickets, lung cancer, night blindness, bronchitis, tuberculosis, stroke, malaria.

2 Define the terms **health**, **disease**, **endemic**, **epidemic** and **pandemic**.

3 Outline the advantages to health that may arise from the Human Genome Project.

Balanced diet

We need to eat a balanced diet which is related to our needs. This must include:

- sufficient energy for our needs provided by the macronutrients (carbohydrates, proteins and fats);

- essential amino acids;

- essential fatty acids (linolenic acid and linoleic acid);

- micronutrients – vitamins and minerals;

- water for replacing the water lost in urine, sweat, breath, faeces;

- fibre for preventing constipation.

Macronutrient requirements

These nutrients are needed for growth and repair of tissues and for making chemicals such as haemoglobin and phospholipids.

Proteins are digested to amino acids. The essential amino acids must be in the diet as the body cannot make them.

Fats provide more energy than the other macronutrients and are used as a long-term energy store in the body. The essential fatty acids cannot be synthesised by cells and must be in the diet. Fats are used to make phospholipids for cell membranes.

Carbohydrates provide energy for respiration (especially for the brain) and are stored as glycogen in liver and muscles.

> ▶ Cells use amino acids to make proteins, such as enzymes and carrier proteins and haemoglobin (see pages 18 and 24, and 14).

> ✓ *Quick check 1*

Dietary Reference Values (DRVs)

A UK government committee surveyed the needs of different groups of people for energy and nutrients. The groups were determined by age and gender. The committee found that requirements differed within each group. The results for each group showed a normal distribution as shown here for vitamin A in 17-year-olds.

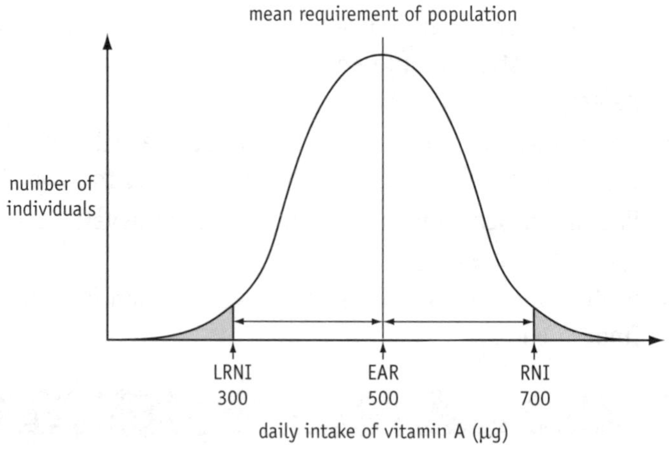

The three DRVs are:

- Reference nutrient intake (RNI);

- Estimated average requirement (EAR);

- Lower reference nutrient intake (LRNI).

Fig. 29.1

These are used to estimate the energy and nutrients required by different groups of people in the UK. The RNIs for protein, vitamins and minerals indicate the amounts that provide sufficient for *most* of the population. People who are taking less than the LRNI may be suffering from a deficiency. The EAR is used for energy. This is because energy requirements vary so much within groups that it is not possible to estimate how much an individual in a particular group needs.

Using DRVs

1 Dietary surveys. DRVs allow nutritionists to assess the diet of groups of people and individuals.

2 Planning diets. Nutritionists develop diets for groups of people, for example those who live in institutions and base these on the DRVs.

✓ Quick check 2

Energy requirements

People's energy requirements are influenced by a number of factors: basal metabolic rate (BMR) – the energy required to maintain the body; physical activity; occupation; age; gender; body mass; growth; (for women) pregnancy and lactation (breast feeding).

Nutrient requirements

The table shows the RNIs for protein and two minerals, calcium and iron.

age	protein (g day^{-1})		calcium (mg day^{-1})		iron (mg day^{-1})	
	males	females	males	females	males	females
0–3 months	12.5	12.5	525	525	1.7	1.7
7–9 months	13.7	13.7	525	525	4.3	4.3
1–3 years	14.5	14.5	350	350	6.9	6.9
7–10 years	28.3	28.3	550	550	8.7	8.7
15–18 years	55.2	45.4	1000	800	11.3	14.8
19–50 years	55.5	45.0	700	700	8.7	14.8
60+ years	53.3	46.5	700	700	8.7	8.7

❂ You may be given tables like this as data in an exam question. You should be able to explain why some groups of people require higher intakes of nutrients such as iron or protein than others.

Points to note:
- **Infancy** (first year of life). Requirements for all three nutrients is high during infancy as children are growing very fast at that stage.
- **Adolescent growth spurt.** Requirements increase to provide energy and materials for growth.
- **Iron**. Females need a greater intake of iron because they lose blood when they menstruate.
- **Calcium** intakes should be maintained in old age to protect bones against degenerative diseases, such as osteoporosis.
- **Pregnancy and lactation**. Intakes of nutrients should increase or be maintained as the body's metabolism works faster and nutrients are supplied to support the growth and metabolism of the fetus and the baby.

✓ Quick check 3

❓ **Quick check questions**

1 State the components of a balanced diet and state the functions of each.

2 Name the three Dietary Reference Values and explain how they are used.

3 Explain why requirements for energy and nutrients increase during pregnancy and lactation.

Malnutrition

Malnutrition means eating much less or much more than needed. People who are starving do not have sufficient energy or nutrients and often show symptoms of protein–energy malnutrition. If the body does not receive enough energy, then basal metabolism slows down and physical activity becomes difficult. The lack of protein means that growth and development slow down considerably. Eating more than is needed can lead to obesity which is associated with many risks to health, such as diabetes and heart disease.

✓ *Quick check 1*

Protein–energy deficiency

Young children are the most susceptible to protein–energy deficiency and they may suffer from a range of symptoms. During starvation, energy reserves in glycogen and fat are used first. When these are exhausted, protein in muscles and other tissues is broken down leading to muscle wasting and an emaciated appearance. The most extreme forms are called **kwashiorkor** and **marasmus**. The table shows some features seen in children suffering from these conditions.

kwashiorkor	marasmus
bloated appearance	very thin with wrinkled skin
'moon face'	'old man's face'
apathetic	mentally alert

Growth and development are halted in children and if this continues for long enough, permanent damage may result and children will not attain their growth potential.

Anorexia nervosa

Anorexia nervosa is
- a mental disease as well as a physical disease;
- often caused by poor body image and problems with adjusting to adulthood;
- more common in girls, but boys suffer from it as well.

People avoid eating with the result that they show many of the symptoms associated with protein–energy malnutrition, such as muscle wasting. There may be limited sexual development and in girls the menstrual cycle may stop, leading to infertility. A deficiency of protein means that the immune system does not function well and anorexics are susceptible to many infectious diseases.

✓ *Quick check 2*

Deficiency of vitamin A

The body uses vitamin A in two main ways.

1 Rod cells in the retina convert vitamin A into a pigment, rhodopsin. This is used to see in light of low intensity.

2 Epithelial cells in the skin and in the lining of the gut and the lungs convert vitamin A into retinoic acid which helps to protect us against invasion by pathogens.

People with insufficient vitamin A in their diet suffer from deficiency diseases:

- night blindness – rods do not make enough rhodopsin so people cannot see in dim light;
- xerophthalmia – the surface of the cornea is scarred, which leads to blindness;
- poor defence against diseases, such as measles.

Those most at risk of vitamin A deficiency are children who eat a diet poor in meat and vegetables.

Deficiency of vitamin D

Vitamin D is absorbed from foods such as eggs and oily fish. It is converted by the liver into an active form that acts as a hormone to stimulate gut cells to absorb calcium and cells in the bones to use calcium to make the hard matter of bone.

If vitamin D is deficient, then bones are not formed properly. There are two deficiency diseases:

- rickets, when the deficiency affects children – the most obvious effect is the bending of the long bones in the legs which bow outwards;
- osteomalacia, when the deficiency occurs in adults – this causes a softening of the bones, pain and weakness.

> Skin cells make vitamin D from cholesterol when exposed to ultra violet light from the sun. Vitamin D is not absorbed from sunlight.

✓ *Quick check 3*

Obesity

A person who is very overweight is obese. Body mass index is a way of determining whether you are overweight. This is calculated as

$$BMI = \frac{\text{body mass in kg}}{\text{height in metres}^2}$$

BMI	category
<20	Underweight
20–25	Acceptable
25–30	Overweight
>30	Obese

> Be careful here about *mass* and *weight*. People talk about weight in kg (or stones and pounds) when in scientific terms, they should say *mass*.

The prevalence of obesity is increasing in affluent countries as people eat far more food than they need. Excess food is stored as fat.

People who are obese are at risk of developing degenerative diseases such as coronary heart disease, diabetes, arthritis, high blood pressure and some forms of cancer.

✓ *Quick check 4*

> *Prevalence* here means the number of people in a population who are obese. In the UK, it is now as high as 18% of the population.

? Quick check questions

1. Explain the term **malnutrition**.
2. Describe the symptoms of anorexia nervosa.
3. State the diseases caused by deficiencies of vitamins A and D.
4. Describe how to determine whether someone is obese.

The lungs and gas exchange

In the lungs, oxygen diffuses from the air in the alveoli into the blood. Carbon dioxide diffuses in the reverse direction. This is gas exchange (see page 26). Ventilation of the alveoli is achieved by moving the diaphragm and the rib cage when we breathe. Air moves through a system of airways to reach the alveoli. Fig. 31.1 shows these airways and their structure when viewed through the light microscope.

✓ Quick check 1

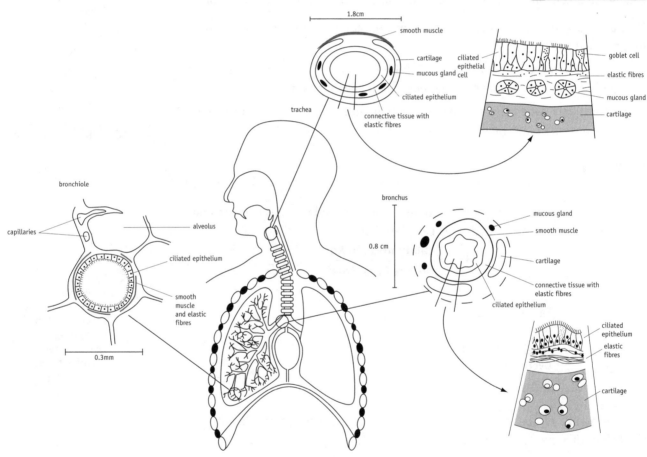

Fig. 31.1

This table shows the distribution of the main tissues and cells in the airways.

tissue / cell	trachea	bronchus	bronchiole	alveolus
cartilage	✓	✓	✗	✗
goblet cells	✓	✓	✗	✗
ciliated cells	✓	✓	✓	✗
smooth muscle	✓	✓	✓	✗
squamous epithelium	✗	✗	✗	✓
elastic fibres	✓	✓	✓	✓

The table shows the functions of cells, tissues and fibres in the gas exchange system.

cartilage	smooth muscle
provides strength to trachea and bronchus; holds open the airways so there is little resistance to air flow.	contracts to narrow the airways.
goblet cells	**elastic fibres**
secrete mucus; mucus is sticky and collects particles of dust, spores, and bacteria.	stretch when breathing in and filling the lungs; recoil when breathing out to help force air out of the lungs.
ciliated cells	**squamous epithelium**
move mucus up the airways towards the mouth.	gives short diffusion pathway for oxygen and carbon dioxide in the alveoli.

✓ *Quick check 2*

Lung volumes

A spirometer may be used to measure lung volumes. Fig. 31.2 shows how a spirometer works.

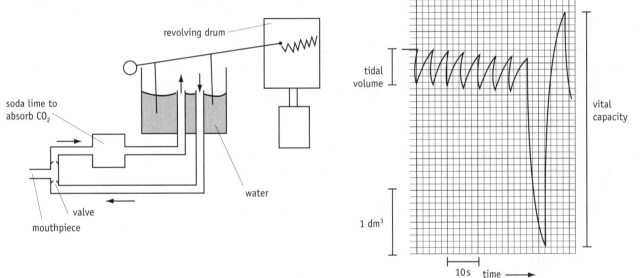

Fig. 31.2

Fig. 31.3

Fig. 31.3 shows a spirometer trace of a 17-year-old male who breathed normally for 2 minutes and then took a deep breath and breathed out as much air from his lungs as possible.

- Tidal volume is the volume breathed into the lungs in one breath. At rest it is usually 500 cm^3.
- Vital capacity is the maximum volume of air that can be breathed out of the lungs after taking a deep breath.

✓ *Quick check 3*

❓ *Quick check questions*

1 Describe the pathway taken by air as it passes from the atmosphere to the alveoli.

2 State the distribution and functions of the following in the gas exchange system **cartilage**, **goblet cells**, **cilia**, **smooth muscle** and **elastic fibres**.

3 Explain what is meant by the terms **tidal volume** and **vital capacity**.

Pulse rates and blood pressure

The heart makes adjustments in order to meet the demands of the body for oxygen and nutrients. This is especially important during exercise. Adjustments are made in the

- heart rate – the number of beats per minute;
- blood pressure – the force exerted on the blood as it is pumped out of the ventricles.

The performance of the heart is monitored by recording changes in the pulse and blood pressure. When the heart contracts, blood flows into the aorta. The force of the blood against the walls of the aorta and other arteries is the systolic pressure which causes the walls to expand slightly. When the heart relaxes, the pressure against the walls of the arteries decreases.

You can feel the expansion of the arteries as blood surges through them when the heart contracts and blood is forced out of the left ventricle. This is the pulse. You can take your own pulse at pressure points – places where an artery is close to the skin. The wrist is the usual place to take the pulse. The pulse rate is a direct measurement of the heart rate.

Pulse rates change during the day and increase with age. Pulse rates are good indicators of aerobic fitness (see page 68). Many endurance athletes have low pulse rates at rest, because they have large hearts that pump large volumes of blood with each beat. This means that their hearts do not need to beat as fast as those of non-athletes.

> You may wish to revise pages 48 to 49 on the heart before reading this section.

> ✓ Quick check 1

Blood pressure

Blood pressure monitors have a cuff that fits around the arm. When the cuff is inflated the pressure becomes greater than that in the main artery in the arm so stopping the blood flow. As the pressure in the cuff is released the monitor detects the highest and lowest blood pressures in the artery.

- Systolic blood pressure – this is the highest blood pressure and is usually about 120 mmHg at rest (15.8 kPa).

- Diastolic blood pressure – this is the lowest blood pressure and is usually about 80 mmHg (10.5 kPa).

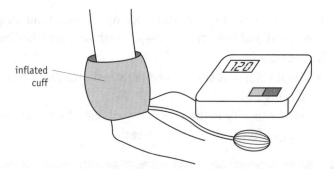

inflated cuff

Fig. 32.1 A digital blood pressure monitor.

During exercise muscles require more energy so respire faster. The heart rate increases so that more blood flows through the muscles to provide oxygen for aerobic respiration.

At the end of exercise, the heart rate slows down, but does not return immediately to the resting rate as muscles still require more oxygen than at rest. The time taken to return to the resting value is the recovery time. During strenuous exercise, the heart rate increases until it reaches the maximum. You can calculate your maximum heart rate by subtracting your age from 220.

You can see from Fig. 32.2 that systolic blood pressure increases during exercise. This helps to deliver oxygenated blood more efficiently to the muscles. Diastolic pressure changes very little during exercise.

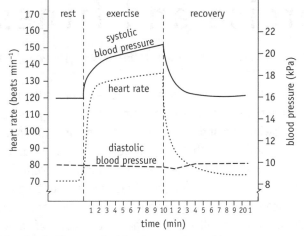

Fig. 32.2 Changes in blood pressure and heart rate.

✓ *Quick check 2*

Blood pressure readings are useful indicators of people's state of health

Hypertension is high blood pressure. In people with high blood pressure, both systolic and diastolic pressures are high. Blood pressure in the population shows a normal distribution, but the values shown in the table are used to define different categories.

category	blood pressure (mmHg)	
	systolic	diastolic
below normal	< 100	< 60
normal	100–139	60–89
borderline	140–159	90–94
hypertension	> 159	> 94

There are several risk factors that may be responsible for causing blood pressure to rise: heredity; race; gender; age; mass; diet; nicotine; heavy alcohol consumption.

 ✓ *Quick check 3*

? *Quick check questions*

1 Explain what is meant by the term **pulse**.

2 Describe how pulse measurements can be used to assess the effect of exercise on the body.

3 Explain the terms **systolic blood pressure**, **diastolic blood pressure** and **hypertension**.

Exercise and fitness

Broadly speaking there are two forms of exercise. There is the type that is undertaken in short bursts of activity, such as weightlifting and sprinting, and there are endurance events such as swimming and long-distance cycling and running. The first type relies upon having strong muscles that are provided with energy mainly by anaerobic respiration. During training for these events, athletes concentrate on strength-building exercises. The second type uses muscles that can work for long periods of time without tiring. Endurance athletes mainly use aerobic respiration and therefore need a good blood supply for efficient delivery of oxygen and removal of carbon dioxide. These athletes have powerful hearts to pump blood efficiently and good ventilation systems for efficient gas exchange.

Oxygen deficit and oxygen debt

At rest, muscle tissue gains energy by aerobic respiration. However, when exercise starts the demand for energy increases steeply and there is not enough oxygen supplied by the blood to support aerobic respiration. It takes time for blood vessels in the muscles to widen and for the heart rate and breathing rate to increase.

> Remember that aerobic respiration uses oxygen; anaerobic does not.

During the first few minutes of exercise, muscles gain much of their energy from anaerobic respiration. Energy is released from glycogen stores in muscle and lactate is produced. Lactate (or lactic acid as it is also known) accumulates in the muscle tissue and diffuses out into the blood capillaries. As exercise continues, blood vessels in the muscle widen and allow more blood to flow through. This brings oxygen and so the rate of aerobic respiration increases. Lactate is carried away in the blood stream.

You can see from Fig. 33.1 that the body is not able to supply enough oxygen early in exercise. This is an oxygen deficit. After exercise has finished, the lactate produced during the first few minutes, must be removed from the circulation and either respired or converted to glycogen and stored. This occurs in the liver. Notice that the consumption of oxygen at the end of exercise takes some time to return to the resting value. The body is repaying an oxygen debt.

> ✓ *Quick check 1*

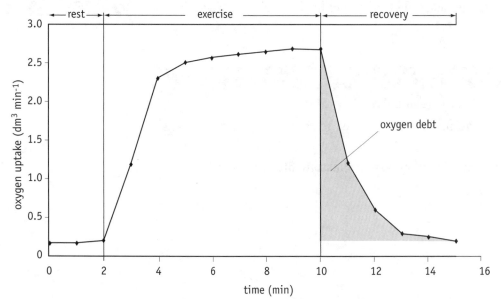

Fig. 33.1

Aerobic exercise and aerobic fitness

Any exercise that uses the cardiovascular system (heart and blood vessels) and the gaseous exchange system (trachea, bronchi and lungs) is described as aerobic. This includes: walking; cycling; jogging; swimming; running; dancing.

Aerobic fitness is a measurement of how effective the heart and lungs are in obtaining oxygen and delivering it to the tissues. It is measured in a variety of ways: resting heart rate; recovery time; oxygen consumption at exhaustion; pulse rates after exercise.

If you wish to improve aerobic fitness, it is recommended that you exercise at 70% of your maximum capacity three times a week. Assessing improvements in aerobic fitness can be carried out before and during a training programme.

✓ Quick check 2, 3

1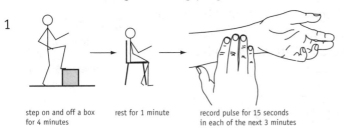

step on and off a box   rest for 1 minute   record pulse for 15 seconds
for 4 minutes   in each of the next 3 minutes

2 Put results in a table and calculate fitness score by multiplying results by 4 and then adding them up to give an estimate of the total number of beats in the 3 minutes

time (min)	pulse per 15 seconds	pulse per minute
1		
2		
3		
	total beats per minute	

3 Caclulate fitness score $\dfrac{24\,000}{\text{total beats}}$ = fitness score

fitness score	rating
<61	poor fitness
61–70	average
71–80	very good
>81	excellent

Fig. 33.2

The benefits of aerobic exercise

Aerobic exercise improves cardiovascular fitness so that the heart and circulation work better at providing oxygen and nutrients to the tissues. The lungs improve and muscles develop better aerobic capacity as the number of mitochondria increase and more enzymes for respiration are made. More capillaries develop inside muscles to improve the supply of oxygen.

? Quick check questions

1 Explain the terms **oxygen deficit** and **oxygen debt**.

2 Describe how you might assess someone's aerobic fitness.

3 Explain how you would find out how much exercise is necessary for an improvement in aerobic fitness.

Smoking and disease

The World Health Organisation considers smoking to be an epidemic. This is because smoking is the cause of numerous diseases and a contributory factor to many more. The major effects are on the gas exchange system (trachea, bronchi and lungs) and the cardiovascular system (heart and blood vessels).

✓ *Quick check 1*

Cigarette smoke

When dried tobacco leaves are burnt they give off a large number of substances that are inhaled into the lungs. Some remain in the lungs; others are absorbed into the blood. This table shows some of these substances and their effects on the body.

Make sure you recognise that 'cardiovascular' refers to heart and blood vessels.

substance	effects on the body
tar	• accumulates in the airways (especially the bronchi); • destroys cilia; • stimulates goblet cells to secrete more mucus; • causes chronic bronchitis and emphysema.
carcinogens	• cause mutations to occur in bronchial epithelial cells leading to formation of tumours (lung cancer).
carbon monoxide	• absorbed into blood; • combines with haemoglobin to form carboxyhaemoglobin; • reduces oxygen carrying-capacity of the blood.
nicotine	• absorbed into the blood; • increases heart rate; • stimulates decrease in blood flow to extremities; • increases the chances of blood clots forming.

✓ *Quick check 2*

Chronic bronchitis, emphysema and lung cancer

disease	changes in the lungs	symptoms
chronic bronchitis	bronchi become obstructed and narrow because: • lining is inflamed • smooth muscle layer thickens • goblet cells and mucous glands produce much mucus	shortness of breath wheezing persistent cough
emphysema	• alveoli become overstretched, lose elasticity and burst • fewer elastic fibres • large gaps in the lungs giving smaller surface area for gaseous exchange	shortness of breath difficulty in breathing out in severe cases people need to breathe oxygen through a mask
lung cancer	• bronchi are blocked by cancerous growths	coughing up blood persistent cough weight loss

✓ *Quick check 3*

Epidemiological evidence

The link between cigarette smoking and lung cancer was first suggested in the 1950s by epidemiologists who collected data from patients with the disease. They found that almost all lung cancer patients were smokers.

Many studies since then have shown that smoking is the major cause of lung cancer. Very few non-smokers develop the disease. Many smokers develop it and die from it. The table below summarises some of this epidemiological evidence

observations	explanation
lung cancer was a rare disease before the 20th century	cigarettes were first made at the end of 19th century; smoking became common early in 20th century
cases of lung cancer increased from 1930s onwards	smoking became common during First World War; it takes 20–30 years for symptoms to develop
more men than women suffer from lung cancer	for most of the 20th century more men than women smoked cigarettes
most people who develop lung cancer are smokers	tar from cigarette smoke contains carcinogens (other causes of lung cancer are very rare)
death rates from lung cancer are highest among people who smoke > 25 cigarettes a day	people who smoke many cigarettes per day expose their lungs to more carcinogens so increasing the chances of cancer

Experimental evidence

There are two main lines of experimental evidence for the link between smoking and lung cancer:

- dogs that were exposed to cigarette smoke in the same way as humans developed cancerous growths in their lungs;

- when substances extracted from tar in cigarette smoke were painted on the skin of mice, tumours started to develop.

These experiments show that cigarette smoke contains carcinogens (cancer-causing agents) that cause genes to mutate so that cells start to divide uncontrollably to give cancerous growths or tumours (see page 37).

✓ *Quick check 4*

? *Quick check questions*

1 Distinguish between the gas exchange and cardiovascular systems.
2 List the four main components of cigarette smoke and state their effects on the body.
3 Describe the symptoms of chronic bronchitis, emphysema and lung cancer.
4 Describe the epidemiological and experimental evidence that confirms the link between cigarette smoking and lung cancer.

Cardiovascular diseases

Diseases of the heart and circulation are cardiovascular diseases which are degenerative diseases brought about by changes that occur in the walls of arteries. When these degenerative changes occur in the coronary arteries of the heart they cause coronary heart disease (CHD). When they occur in arteries in the brain they cause a stroke. CHD and stroke often occur as a result of two events:

- a build up of fatty material known as plaque inside the wall of arteries, and
- a blood clot.

Plaque and blood clots stop the flow of blood to a region of the heart or brain leading to death of surrounding tissues because they do not gain sufficient oxygen and nutrients. This may be fatal.

Note that these events happen in arteries, not in veins.

Atherosclerosis

Atherosclerosis is the progressive build-up of plaque. The flow chart in Fig. 35.1 shows the events that often lead up to the formation of plaque in the lining of artery walls. Plaque enlarges the wall so that there is less space for blood to flow. It also roughens the lining of arteries so increasing the chances of blood clots forming.

Coronary heart disease

Oxygenated blood flows from the aorta into the coronary arteries which supply the muscle in the atria and ventricles of the heart (see page 48). If these arteries become filled with plaque, the heart muscle may be deprived of oxygen and become fatigued. This means that people with this condition find even mild exercise difficult and have chest pains which disappear when they stop exercising. This form of CHD is angina. If a blood clot forms in a coronary artery and blood flow is cut off the heart muscle may die causing a heart attack.

Stroke

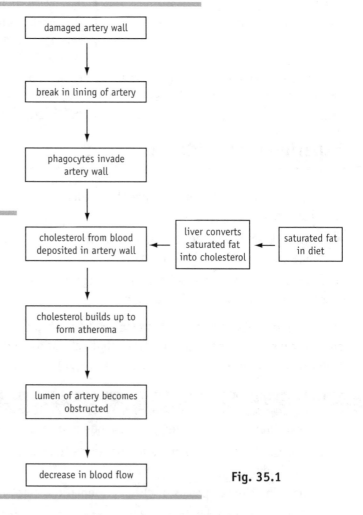

Fig. 35.1

Arteries in the brain may become weak as a result of plaque formation and burst. Blood may also clot within an artery. Brain cells may die and areas of the brain may cease to function. If this is not fatal, then certain functions, such as memory or control of part of the body, may be impaired or lost.

✔ *Quick check 1, 2*

Risk factors

CHD and stroke are multifactorial – many factors contribute to the development of these diseases such as diet, heredity, etc. Diet may be an important factor. The incidence of CHD is high in some countries, for example Finland and the UK, where people consume large quantities of animal fats rich in saturated fatty acids. This stimulates the liver to make more cholesterol, which is transported in the blood to various places in the body such as artery walls. Countries with low consumption of animal fat, such as Japan, have a very low incidence of CHD (see Fig. 35.2).

Global distribution of CHD

Fig. 35.2 shows the changes in death rates among men between the ages of 33 and 74 who are most at risk of early death from CHD. The data is for selected countries with high and low death rates from the diseases.

✓ Quick check 3

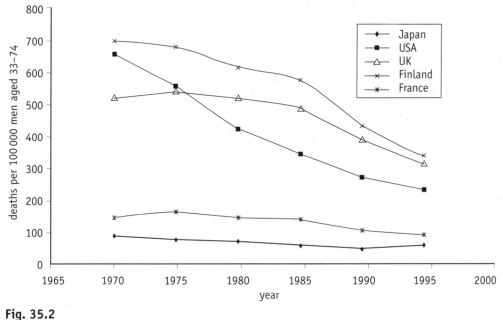

Fig. 35.2

Preventing CHD

Screening the population for the risk factors listed above can help to identify those people most at risk of developing CHD. People who have an inherited disease that raises their blood cholesterol concentration are the highest priority. People with high blood pressure are also a priority. People can do something about some of the risk factors listed above. They can reduce their intake of saturated fats, take more exercise, stop smoking and follow medical advice to lower blood pressure.

Preventing CHD is much cheaper than treating it. Drugs to lower blood cholesterol and lower blood pressure are expensive. Operations, such as coronary by-passes and heart transplants, are expensive and risky. In 1996, the UK government spent £1 630 million on CHD: 32% of that was spent on drugs, 54% on operations and 1% on prevention.

? Quick check questions

1 Explain the terms **cardiovascular disease**, **coronary heart disease**, **stroke** and **atherosclerosis**.

2 Describe the events that occur in the formation of a plaque in an artery.

3 Describe the global distribution of CHD and explain how diet may be responsible for the distribution you describe.

Infectious diseases

Pathogens are organisms that invade the body, multiply in tissues or inside cells and cause disease. Disease transmission is the transfer of a pathogen from infected to uninfected people.

- Malaria and tuberculosis (TB) are caused by pathogens that invade our cells and then spread through the tissues.

- Cholera bacteria invade the gut and cause diarrhoea with the loss of much fluid leading to severe dehydration.

- Human immunodeficiency virus (HIV) infection can lie dormant in T lymphocytes (see page 78) in the body for a long time, but eventually weakens the immune system so that people become susceptible to several opportunistic diseases, such as pneumonia and certain rare cancers. The collection of these diseases is known as AIDS (**acquired immune deficiency syndrome**).

> Make sure you know about methods of transmission and how this helps us to develop methods to control these diseases.

Causative organisms and means of transmission

disease	causative organism (pathogen)	main methods of transmission
cholera	bacterium: _Vibrio cholerae_	bacteria passed out in faeces of infected people contaminate drinking water and food
malaria	protoctist: several species of _Plasmodium_	insect vector female _Anopheles_ mosquito
TB	bacterium: _Mycobacterium tuberculosis_ _M. bovis_	airborne droplets of water _M. bovis_ in milk and meat from infected cattle
HIV/AIDS	virus: Human immunodeficiency virus	• during sexual intercourse • infected blood and blood products • sharing or reusing hypodermic needles • across placenta from mother to fetus

✓ *Quick check 1,2,3*

Worldwide distribution

disease	worldwide distribution
cholera	_associated with displacement of people as a result of war and natural disasters_. West and East Africa, Afghanistan.
malaria	widely distributed throughout the tropics and sub-tropics.
TB	worldwide – throughout developing world, countries of old Soviet Union, among homeless and poor in inner cities in developed world. Especially amongst immigrants from developing countries.
HIV	worldwide: highest prevalence in Africa and South-east Asia.

Control and prevention of infectious diseases

Measures can be taken to control and prevent the spread of disease. This is done by breaking the transmission from infected to uninfected people.

disease	control measures
cholera	<ul><li>good sanitation – provide hygienic removal and treatment of human faeces;</li><li>ensure that drinking water supply is not contaminated by sewage;</li><li>provide clean drinking water which is treated to kill bacteria (e.g. by chlorination);</li><li>fast treatment of people infected with cholera by rehydration therapy (salts and glucose in sterile water) and, in severe cases, with antibiotics.</li></ul>
malaria	<ul><li>control mosquitoes by destroying breeding areas (e.g. draining marshes) and spraying insecticides;</li><li>prevent mosquitoes biting at night by using sleeping nets (most effective when nets are soaked in insecticide every 6 months);</li><li>use drugs which prevent *Plasmodium* spreading through the body;</li><li>use drugs to treat people with malaria to reduce reservoir of infection.</li></ul>
TB	<ul><li>isolation of people during infective stage;</li><li>use antibiotics to treat infected people (course of treatment can be 6–12 months);</li><li>contact tracing to find other people likely to be infected;</li><li>vaccination with BCG vaccine (see page 82);</li><li>TB testing of cattle – destroying any cattle infected with TB;</li><li>pasteurisation of milk.</li></ul>
HIV/AIDS	<ul><li>using condoms or femidoms during sexual intercourse;</li><li>health education about 'safer sex';</li><li>contact tracing to find people likely to be infected;</li><li>blood donations screened for HIV;</li><li>blood donations and blood products heat-treated to kill viruses.</li></ul>

✓ Quick check 4

? Quick check questions

1 Name the organisms that cause cholera, malaria and tuberculosis.

2 Explain how these three diseases are transmitted from an infected to an uninfected person.

3 Explain how HIV infection leads to AIDS.

4 Describe how the spread of cholera, malaria, TB and HIV/AIDS may be controlled.

Antibiotics and the control of infectious diseases

Antibiotics are drugs that interfere with the metabolism of pathogens, such as bacteria and fungi, without interfering with our metabolism. Some antibiotics kill pathogens, others stop their growth, making it possible for our immune system to remove them more easily. Some antibiotics are made by microorganisms, such as fungi. Some antibiotics produced like this are modified chemically after production. Others, such as isoniazid, used to treat people with TB, are synthesised chemically and not by an organism.

✓ **Quick check 1**

Antibiotics are most effective against bacterial diseases. There are very few drugs that can be used to treat viral infections.

Antibiotic resistance

This is a serious problem. This is because antibiotics lead to the death of all genotypes of bacteria that are susceptible, leaving any that are resistant. Resistance arises in the first place by mutation. If an antibiotic is widely used, it is highly likely that resistant pathogens will appear.

The widespread use of antibiotics over the past 50 years has led to the appearance of resistant strains of TB. Some strains of *Mycobacterium tuberculosis* are resistant to several antibiotics and are known as multi-drug resistant TB or MDR-TB. These represent a great threat to health as it is difficult to treat them with antibiotics.

▶ You may have learnt about antibiotic resistance in your GCSE course – it is an example of natural selection.

▶ Pathogens become resistant to antibiotics, not immune to them.

To reduce the chances of resistance the following steps should be taken.
- Only prescribe antibiotics when necessary.
- Change antibiotics so that the same one is not used continuously.
- Keep some antibiotics in reserve for use when all others fail.
- Test pathogens from patients to find the most appropriate antibiotics to use.
- Ensure that patients take the prescribed dose every day and complete the course of treatment.

✓ **Quick check 2**

Biological, social and economic factors in controlling infectious disease

Cholera

- The disease is rare in countries with efficient sewage disposal and the provision of clean, piped drinking water; some developing countries cannot afford to provide these facilities so the risks of cholera and other diarrhoeal diseases are high;
- carriers of the disease may infect others when they prepare food without washing their hands after defecating;
- cholera is common when people have to leave their homes during wars, civil disturbance or natural disasters, such as earthquakes and floods, and live in refugee camps or makeshift accommodation;

- rehydration therapy, if given within 24 hours of developing the disease, cures most cases – this is not always available fast enough;
- a successful vaccine is not widely available to provide immunity to cholera (see page 82).

Tuberculosis

- TB is controlled by using the BCG vaccination – this reduces the chances of people developing the disease;
- the effectiveness of the BCG vaccine varies – it is not successful in providing immunity in some countries, e.g. India;
- people living in poverty are most at risk of developing TB because they tend to live in overcrowded conditions which favour the spread of the bacterium;
- homeless people are also at high risk as they often sleep close together in shelters or large dormitories;
- the drug treatment only successfully destroys *Mycobacterium* in the lungs if taken for 6 months to a year – many people with the disease stop taking their drugs when they begin to feel better, but before the bacteria in their lungs are all killed;
- many people infected with TB live in developing countries that do not have the medical facilities to maintain contact with patients until they complete their course of drugs;
- schemes, such as DOTS (Direct Observation Treatment Short Course) developed by the World Health Organisation, are achieving success in combating TB by making sure the patients are reminded to take their drugs and finish their course of treatment. Patients are watched to ensure that they take their drugs.

Malaria

- the disease was eradicated from many parts of the world in the 1950s by using insecticides to destroy mosquitoes;
- mosquitoes are resistant to many insecticides;
- *Plasmodium* is resistant to some of the drugs used to control it;
- there is no successful vaccine widely available as yet;
- political unrest and warfare have prevented the adoption of effective measures of control.

HIV/AIDS

- many drugs used in combination can delay the onset of AIDS;
- these drugs are expensive and not available to people in developing countries;
- there is no cure and no vaccine as yet;
- health education has had a limited success in changing people's sexual practices and in preventing drug takers sharing needles.

✔ *Quick check 3*

Quick check questions

1 Explain how antibiotics act to control infectious diseases.
2 State the problem that arises when one antibiotic is used widely.
3 Outline the problems involved in controlling the spread of cholera, malaria, TB and HIV infection.

The immune system

The body's defences against infectious diseases make up the **immune system**. In this section we deal with the cells of the immune system. On pages 80 and 81, we will look at the structure and function of antibodies. There are four main groups of cells in the immune system.

cell type	distribution	function
neutrophil	blood and tissues	phagocytosis
macrophage	tissues, e.g. lungs	phagocytosis
B lymphocyte (B cell)	blood and lymph nodes	production of antibodies
T helper lymphocyte (Th cell)	blood and lymph nodes	stimulate B cells to divide and produce antibodies; stimulate phagocytosis
T killer lymphocyte (Tk cell)	blood and lymph nodes	destroy cells infected with viruses

Fig. 38.1 shows where these cells originate in the body and where they mature. All the cell types originate from groups of stem cells in bone marrow. As they mature, B and T lymphocytes gain glycoprotein receptors on their cell membranes. B lymphocytes mature in the bone marrow, T lymphocytes mature in the thymus. There is more about these receptors on page 79. Macrophages spread throughout the body. They are in organs, such as the lungs, skin, liver and gut.

Fig. 38.2 shows how a neutrophil destroys some bacteria.

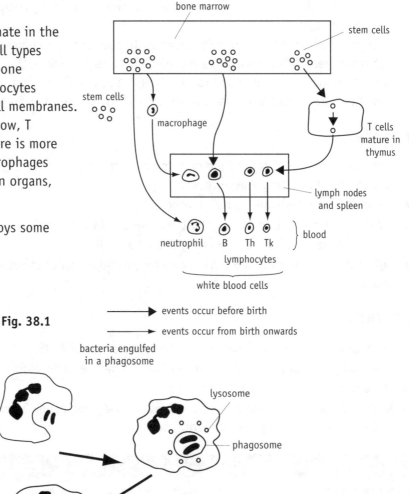

Fig. 38.1

Fig. 38.2

Phagocytes, such as neutrophils and macrophages, are not very successful at removing invading bacteria. They are much more effective when working alongside lymphocytes. The events that occur during an **immune response** make a more successful defence against pathogens.

> ❶ Antigens are certain foreign substances that invade the body stimulating the production of antibodies.

The immune response

Throughout the body, especially in lymph nodes, and organs like the spleen, there are many millions of B and T lymphocytes. On the surface of these cells are glycoproteins that recognise antigens on the surfaces of specific pathogens. When a pathogen enters the body, the small number of B and T cells with glycoproteins that are able to bind to antigens on the pathogen are selected to respond. The selected cells have glycoproteins that are complementary in shape to the antigens. These selected cells divide by mitosis and form large numbers of identical cells. Some cells become effector cells; others become **memory** cells. This happens to both B and T cells. B cells that become effector cells (plasma cells) make and release large quantities of antibodies.

> ❶ Antibodies are proteins – make sure you know that their tertiary structure gives them specific shapes so that they can bind tightly to antigens. This is like enzymes and substrates – see page 18.

Fig. 38.3 shows what happens during an immune response. Note this is a diagram. Antibody molecules are very much smaller than the cells that secrete them.

The events that are shown in Fig. 38.3 occur during the first response (primary response) to an invasion by a pathogen. When a pathogen with the same antigens invades again, the secondary response is much faster. Antibodies are produced very quickly because the large numbers of **memory cells** divide and become effector cells within a short period of time. This means that the pathogen is destroyed before it can cause disease.

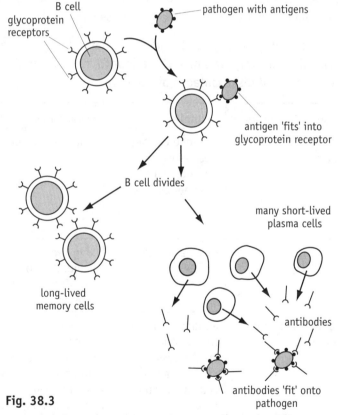

Fig. 38.3

✓ *Quick check 2, 3*

❓ **Quick check questions**

1 Describe what happens when a neutrophil carries out phagocytosis.

2 Define the following terms: **immune system**, **immune response** and **antigen**.

3 Make a table to compare B and T cells. Use the following headings: origin, site of maturation, action during an immune response.

Immunity and antibodies

People are immune to a disease when they can mount a fast, effective defence against a pathogen so that they do not develop any symptoms of the disease concerned. People can become immune to a disease in two different ways.

Active immunity

The immune system develops antibodies directed against the specific pathogen involved. Active immunity is usually long-term as memory cells are produced.

Passive immunity

The immune system does not develop antibodies of its own. Antibodies are transferred from another source. Memory cells are not produced in passive immunity so the immunity can only be short-term, several weeks or months at most.

Both types of immunity can be gained in natural ways and artificially. The table shows different ways in which these four forms of immunity are gained and the advantages and disadvantages of each type.

> **Active** – antigens enter the body and stimulate an immune response. **Passive** – antigens do not enter the body; antibodies come from somewhere else; there is no immune response.

type of immunity	example	advantages	disadvantages
active immunity (natural)	someone catches measles which promotes an immune response	long-term immunity	immune response takes time; protection not immediate; symptoms develop; disease may be fatal
active immunity (artificial)	vaccination against measles	long-term immunity; no need to suffer from the disease	immune response takes time; protection not immediate
passive immunity (natural)	antibodies passed from mother to child across placenta and in colostrum (breast milk)	immediate protection to common diseases that the mother has had or been vaccinated against	short-term; antibodies are destroyed; no memory cells produced
passive immunity (artificial)	antibodies against tetanus toxin collected from blood donations and injected	immediate protection to specific disease, e.g. tetanus, diphtheria	short-term; antibodies are destroyed; no memory cells produced

Someone can become immune by being vaccinated (artificial active) or by being injected with antibodies (artificial passive). Someone involved in an accident who has open wounds may be given both types of immunity to protect them against tetanus.

Antibodies

Antibodies are glycoprotein molecules that are made by B cells that develop into effector cells during an immune response. These effector cells are called **plasma cells**.

Antibodies bind to large molecules that invade the body. These large molecules are called antigens. Antigens are on the surfaces of pathogens; sometimes they can be molecules released by pathogens such as the toxin released by tetanus bacteria.

Antibodies are made whenever an antigen enters the body. The first immune response to a specific antigen is the **primary response**. This is slow because there are few B cells able to make the necessary antibody. Whenever the same antigen enters the body a **secondary response** occurs. This produces large quantities of the specific antibody because there is a large clone of memory cells which develop into plasma cells very quickly. The graph in Fig. 39.2 shows this.

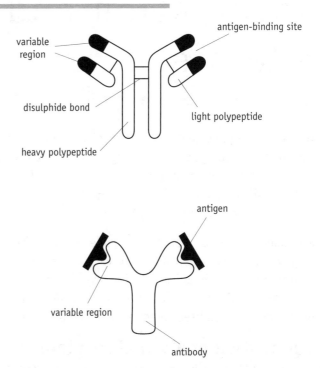

Fig. 39.1 (a) Simplified structure of a typical antibody molecule. (b) How its structure allows it to bind to its specific antigen.

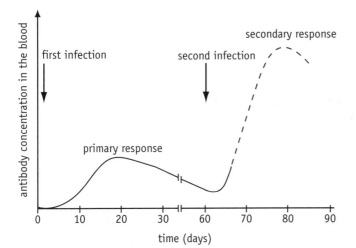

Fig. 39.2

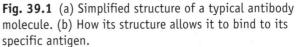 Make sure that you can explain why immunity to one disease does not give immunity to other diseases. Think about what happens during the immune response.

✓ *Quick check 1*

? *Quick check questions*

1 Distinguish between the following terms: **antigen** and **antibody**; **passive** and **active immunity**; **natural** and **artificial immunity**; **primary** and **secondary immune responses**.

Vaccination and allergies

We saw on page 80 that active immunity may be gained artificially by vaccination. A vaccine is a preparation of an antigen or many antigens for a specific infectious disease and is injected or given by mouth. The BCG vaccine for TB is an attenuated (weakened) live form of *Mycobacterium bovis*, the pathogen that causes the disease in cattle and can be caught by humans.

A vaccine stimulates a primary immune response (see Fig. 39.2) and so it takes time to gain immunity. The response to vaccines is good if the vaccine is a 'live' vaccine containing living organisms. To improve the response to other vaccines, boosters are given to stimulate secondary responses and the development of larger clones of lymphocytes specialised to act against the pathogen concerned.

✔ **Quick check 1**

People are advised to have boosters for tetanus every 10 years to maintain their protection. Make sure you can explain why in terms of the secondary response.

Vaccination eradicated smallpox

Smallpox was the first disease to be eradicated from the world. It is likely that polio will be the next. Smallpox was a terrible disease caused by a virus; it was suitable for eradication because:

- there was only one strain of the pathogen;
- the virus did not infect animals;
- diagnosis of the disease was easy;
- any person who became infected with the virus developed symptoms of the disease.

This meant that there was no reservoir of viruses that could reinfect people. The vaccination programme, coordinated by the World Health Organisation, was successful because:

- a 'live' vaccine was used which meant that boosters were not necessary;
- the vaccine was freeze dried so was suitable for use in the tropics;
- only one vaccine was necessary and so was cheap to produce.

✔ **Quick check 2**

Eradication of other diseases

Other diseases are more difficult to eradicate. Some of the reasons for this are:

- some pathogens exist in many strains which keep changing by mutation (e.g. influenza);
- the pathogens also live in animals, e.g. malaria and mosquitoes;
- the pathogens invade the human gut where the immune system does not work very efficiently, e.g. cholera.

It is difficult to develop vaccines against diseases that are caused by eukaryotic organisms, such as *Plasmodium*, because they have many genes that code for cell surface antigens. Different strains of *Plasmodium* have different antigens. As

Plasmodium passes through its different stages in liver and blood cells, it expresses different antigens. It remains inside cells which makes it difficult for antibodies to be effective. It is difficult to organise vaccination programmes for measles, which primarily affects young children. Often, in developing countries, it is not possible to vaccinate them soon enough after birth to give them protection when they are vulnerable to infection.

✓ *Quick check 3*

Allergies

Allergies, such as hay fever and asthma, are caused by the immune system over reacting to harmless antigens known as allergens. Some allergens, such as pollen grains and the house dust mite and its faeces, are antigenic because they are covered by large molecules. Some people become sensitised to these substances during a primary immune response. When the same substances are encountered again, there is an exaggerated secondary response, which can be serious.

Fig. 40.1 shows how people become sensitised to allergens and what happens on any subsequent exposure.

The allergen for hay fever is most often grass pollen. Hay fever is a very unpleasant allergic reaction which leads to inflammation of the nose, eyes and throat. However, it is seasonal only lasting as long as grass plants are in flower and releasing pollen into the air. Asthmatics may suffer attacks at any time. Inflammation occurs in the airways, such as the bronchi, which become filled with mucus and fluid leaking from the blood. Often, the muscles surrounding the airways contract as well. All this increases the resistance to air flow and asthmatics have great trouble in breathing.

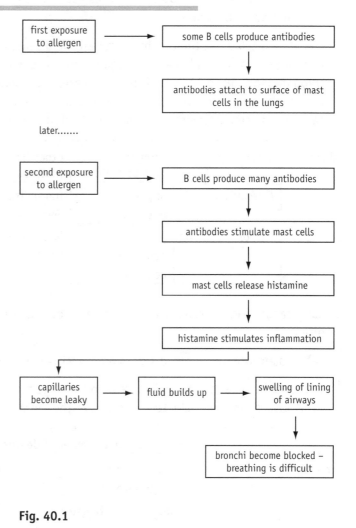

Fig. 40.1

✓ *Quick check 4*

? *Quick check questions*

1 Explain what is meant by a **vaccine**.
2 Outline the reasons for the eradication of smallpox.
3 Explain why it is proving difficult to eradicate cholera, malaria, TB and measles.
4 Describe how asthma develops.

End-of-module questions

In the examination for this module you could be asked to compare features of different diseases by completing a table. The first question is designed to help you with your revision of this module as it gives you a check list of all the diseases you should know something about.

1 Complete a table like the one below for all the diseases listed. Information about two diseases has already been entered to show you what to do. Look back through pages 58 to 83 to find the information you need. Also look at a copy of the specification for this module. You may also have other books to which you can refer. Keep a note of useful sources of information in the last column.

disease	cause	categories (see page 58)	other important features	where to find more information
malaria	*Plasmodium*	physical, infectious	transmitted by mosquitoes	Revise Biology pages 74, 77
rickets	lack of vitamin D	physical, non-infectious, deficiency	bones do not harden	Revise Biology page 63

▶ Diseases that you should know about: lung cancer, chronic obstructive pulmonary disease (chronic bronchitis and emphysema), coronary heart disease, stroke, obesity, anorexia nervosa, night blindness, rickets, osteomalacia, kwashiorkor/marasmus, tuberculosis, malaria, cholera, HIV/AIDS, smallpox (now eradicated), measles, asthma.

2 Lung cancer and coronary heart disease are both classified as degenerative diseases.

(a) Explain why these two diseases are classified as degenerative diseases. [3]

The table below shows the estimated mortality from lung cancer and coronary heart disease in some selected countries in the 1990s.

▶ Questions 2–7 are of the type you might find in the examination for this module.

country	annual number of deaths per 100 000			
	lung cancer (all ages)		coronary heart disease (people aged 35–74)	
	males	females	males	females
Finland	42	7	340	93
United Kingdom	47	21	314	114
United States	52	27	224	92
Japan	31	9	60	23
France	47	6	92	22

▶ Study data like this carefully before you read the questions. Make sure you can identify any trends.

(b) Explain why the death rates in the table are expressed 'per 100 000'. [1]

(c) With reference to the table, describe the differences between death rates for men and women in the selected countries. [4]

(d) Suggest **two** types of epidemiological data that could be collected in these countries to see if there is any link between smoking and heart disease [2]

(e) Outline the effects of long-term smoking on the cardiovascular system. [8]

▶ Questions 2(e), 3(d), 4(c) and 6(c) all require an extended answer and there will be one mark for the way in which you write your answer.

3 Protein in the diet provides essential amino acids and energy.

(a) Describe the role of essential amino acids in the body. [2]

(b) List **three** organic compounds, other than proteins, that provide energy in the diet. [3]

(c) State **three** factors that influence people's requirement for energy in the diet. [3]

(d) A young child does not receive enough energy in the diet over a long period of time. Explain the likely effects of this on the child's body. [7]

4 (a) Describe how the gaseous exchange surface in the lungs is protected. [4]

(b) Describe the functions of cartilage and elastic fibres in the lungs. [4]

(c) Outline the adjustments that occur in the body at the beginning of a period of strenuous exercise. [8]

5 (a) State **three** advantages for health of taking *aerobic exercise*. [3]

A person follows a training programme to improve their aerobic fitness.

(b) Describe how improvements in aerobic fitness could be measured during such a training programme. [4]

6 (a) State **three** ways in which HIV infection is transmitted. [3]

(b) Explain why it is proving very difficult to control the spread of HIV through the world's population. [4]

(c) Explain why infectious diseases, such as malaria and cholera, are much more common in developing countries than in developed countries. [8]

7 The measles virus enters a person's body. As part of the immune response to the disease, some B lymphocytes divide to form plasma cells that secrete antibodies specific to the measles virus.

(a) Describe how the structure of antibodies makes them specific to the measles virus. [4]

(b) Explain why this immune response to measles is an example of active immunity and not passive immunity. [3]

(c) Explain why vaccination is considered to be *artificial active immunity*. [2]

(d) A worldwide vaccination programme for measles has not yet eradicated the disease. Discuss the reasons why measles has so far not been eradicated. [4]

Answers to Quick Check Questions

Many of the quick check questions are designed to test your understanding and recall of material in the spreads. You will find many of the answers quite easily. However, some are more difficult and require you to put several ideas together. Answers to selected quick check questions are given here.

Module: Biology Foundation

Cell structure (page 4)

1 High resolution, so that details of cell structure can be seen e.g. membranes, organelles such as mitochondria, and endoplasmic reticulum.

2

feature	plant cell	animal cell	prokaryote
cell wall	✓	✗	✓
nucleus, nucleolus	✓	✓	✗
mitochondria	✓	✓	✗
vacuoles	few, large	if present, small	✗
Golgi body, ER	✓	✓	✗
chloroplast	✓	✗	✗

3 See the table on page 5. One example: photosynthesis occurs in chloroplasts.

Tissues and organs (page 6)

1 Movement of organisms (e.g. *Stentor*); move fluids (e.g. mucus in lungs); move eggs along fallopian tubes.

Biological molecules 1 – carbohydrates (page 8)

1 −OH on carbon 1 is above the ring in β glucose, below the ring in α.

2 During the reaction between −OH groups on carbon atom 1 and carbon atom 4, water is formed leaving an oxygen 'bridge' between the two glucose units.

3 These molecules are insoluble, compact and can be broken down into many molecules of glucose.

4 Cellulose has many projecting −OH groups. These form hydrogen bonds with other molecules of cellulose; these bonds give cellulose great strength to resist the pressure exerted by plant cell vacuoles on cell walls.

Biological molecules 2 – lipids (page 10)

2 Different fatty acids, which can be saturated, monounsaturated or polyunsaturated.

3 Phospholipids have two fatty acids, triglycerides have three; phospholipids have a phosphate group and another hydrophilic group (e.g. choline), triglycerides do not.

4 Hydrophilic 'heads' of phospholipids are attracted to water, but hydrophobic 'tails' are not. Two layers of phospholipids form a bilayer in water with a hydrophobic core from which water is excluded.

Biological molecules 3 – proteins (page 12)

2 *Primary* – the sequence of amino acids.

Secondary – polypeptide is folded into an α-helix or a β-pleated sheet.

Tertiary – complex folding of the secondary structure to give a specific three-dimensional shape.

3 Bonds between amino acids stabilise the tertiary structure; these are hydrogen bonds, disulphide bonds, ionic bonds and hydrophobic interactions. Amino acids that are far apart in the primary sequence are held close together by these bonds in the tertiary structure. This gives rise to specific shapes.

Biological molecules – 4 globular and fibrous proteins (page 14)

1 1 120 000 000

2 Haemoglobin is made of more than one polypeptide (it is made of four).

4

feature	haemoglobin	collagen
type of protein	globular	fibrous
function	transports oxygen and carbon dioxide	strengthens tissues
number of polypeptides	4	3
shape of molecule	complex 3D shape	triple helix

Water and ions (page 16)

3 Advantages: temperature does not change quickly; little danger of dehydrating; water provides support. Disadvantages: water is dense, so much energy used in movement through it; not much oxygen is dissolved in it.

Enzymes (page 18)

2 Provides an active site where reaction can occur. In the active site substrate molecules are brought close together so there is more chance of them reacting.

3 This should show a substrate molecule with a shape *complementary* to the active site. The substrate should be shown moving into the active site, inside the active site and then leaving as two product molecules.

Experiments with enzymes (page 20)

2 At the beginning there is a high concentration of substrate molecules, so that there are many collisions between enzyme molecules and substrate molecules. With time, concentration of substrate decreases so there are fewer collisions.

3 In both (a) and (b), the reactions would occur faster.

Cell membranes (page 24)

4 Cell surface membrane, nuclear membrane, endoplasmic reticulum, Golgi body, lysosomes, mitochondria and (in plants) membrane around large vacuole, chloroplasts.

Exchanges across membranes (page 26)

2 Both molecules are polar, but carbon dioxide is much smaller and can pass between phospholipids, glucose is too large to be able to do this.

3 Membrane surrounding vacuole (or vesicle) fuses with the cell surface membrane. Contents of vacuole pass out of cell.

4 (i) Squamous epithelium is thin so giving a short distance for diffusion in and out of the blood; (ii) mitochondria provide energy for active transport of ions; (iii) channels are filled with water, so water soluble substances, such as glucose and ions, can pass through.

Nucleic acids – DNA and RNA (page 28)

4

feature	DNA	RNA
type of sugar	deoxyribose	ribose
bases	A, T, C, G.	A, U, C, G.
size	very large	smaller
number of polynucleotides	2 (double helix)	1 (variety of shapes)

Replication of DNA (page 30)

2 Bacteria grown in a medium with a heavy isotope of nitrogen make 'heavy' nucleotides that are incorporated into DNA. When transferred to a medium with light isotope of nitrogen, they make 'light' nucleotides. 'Heavy' DNA settles lower after being spun in a centrifuge than 'light' DNA. After transfer to the light medium, bacteria replicate their

DNA once. When this DNA is extracted from the bacteria and spun in a centrifuge, it is settles midway between the 'light' and 'heavy' forms showing that it is made of 'light' and 'heavy' polynucleotides.

Protein synthesis (page 32)

1 DNA provides a sequence of bases that code for the *primary structure* of the polypeptide. RNA polymerase matches free RNA nucleotides against coding template of DNA to make messenger RNA (transcription). tRNA molecules carry amino acids to ribosomes. The anticodons on tRNA molecules *identify* amino acids in terms of base triplets. On the ribosome anticodons (on tRNA) pair with codons (on mRNA). Amino acids are joined by peptide bonds (translation).

2 tRNA anticodons: CCU; AGC, UUC.

3 There are no anticodons for the three stop codons. Each of these codons signifies the end of a sequence of bases coding for a polypeptide.

Chromosomes and genes (page 34)

1 These are chromosomes that are the same size, have the same shape with the centromere in the same place; they carry the same *genes*.

Nuclear and cell division (page 36)

1 Chromosome condenses during **prophase**. Becomes visible as a double structure – with two chromatids joined at centromere. Attached to the spindle during **metaphase**. Chromatids pulled apart in **anaphase**. Each chromatid uncoils in **telophase**. *When mitosis is completed*, replication occurs during **interphase** to form two molecules of DNA which condense in next **prophase** to form two chromatids.

2 So each chromosome has two *identical* chromatids to transfer to new cells.

4 To halve the number of chromosomes and give haploid nuclei.

Energy and ecosystems (page 38)

1 In coral reef: sunlight is trapped by algae (and other **producers**); **primary consumers** (herbivores) feed on algae; **secondary consumers** (carnivores) feed on primary consumers. Energy also flows in dead matter from producers and consumers to **decomposers**.

2 Decomposers (bacteria and fungi) use enzymes to digest protein to amino acids; amino acids are absorbed; decomposers remove amino groups ($-NH_2$) from amino acids (deamination) and excrete them as ammonium ions (NH_4^+); nitrifying bacteria (see Q.3) convert ammonium ions to nitrate ions that are absorbed by plants.

3 *Rhizobium* absorbs nitrogen gas (N_2) and reduces it to ammonium ions. This is nitrogen fixation. *Nitrosomonas* and *Nitrobacter* convert ammonium ions to nitrate (see page 39 for details).

Module: Transport

Transport systems in animals (page 42)

1 x 16; x4300; x3.5.

2

feature	artery	capillary	vein
thickness of wall	thick	very thin	thin
composition of wall	smooth muscle, elastic fibres, collagen	endothelium (one cell thick)	as artery – but much less
valves	✗	✗	✓
blood pressure	high	low	low
function	transport blood *from the heart*; maintains blood pressure	exchange of substances between blood and tissues	transport blood *to the heart*; valves prevent backflow

Blood, tissue fluid and lymph (page 44)

1 x 2570.

Haemoglobin and gas transport (page 46)

1 Haemoglobin has four haem groups, each of which can combine with a molecule of oxygen. Carbon dioxide combines with free amino groups ($-NH_2$) at the ends of the polypeptides in haemoglobin.

2 Haemoglobin becomes fully loaded with oxygen in the lungs to form oxyhaemoglobin. In tissues, oxyhaemoglobin dissociates to give up oxygen. Quite *small* changes in partial pressure of oxygen in the tissues lead to a *large* decrease in oxygen carried by haemoglobin.

3 Dissociation curve moves to the *right* with increasing partial pressure of carbon dioxide. Carbon dioxide reacts with water inside red blood cells to form hydrogencarbonate ions and hydrogen ions. Oxyhaemoglobin is sensitive to hydrogen ions and dissociates to release oxygen.

4 To increase the capacity to carry oxygen in the blood. The partial pressure of oxygen is low and

therefore saturation of haemoglobin is lower than at sea level. More blood cells are needed to compensate for this.

The heart – structure and function (page 48)

3 0.8 s; 75.

4 The valve closes at 1 because the pressure of blood in the ventricle is greater than in the atrium; it opens at 4 because the ventricle is empty and the pressure in the atrium is higher.

5 The valve opens at 2 because the pressure of blood in the ventricle is greater than in the aorta; it closes at 3 because the ventricle is empty and the pressure in the aorta is higher than in the ventricle.

Transport in plants (page 50)

1 Everything in the xylem moves in the same direction at the same time.

2 The loss of water vapour from the surfaces of plants (e.g. leaves) to the atmosphere.

4 Hollow and without end or cross walls to allow uninterrupted flow of water; supported by lignin to prevent inward collapse when rates of transpiration are high and water columns are under tension.

Transpiration (page 52)

1 e.g. cut stem under water to prevent air locks in xylem; leave to adjust to conditions; keep all conditions constant apart from wind speed; take replicates.

Transport in the phloem (page 54)

3 Source is photosynthesising mesophyll cell; sucrose passes to companion cell via plasmodesmata or along cell walls; companion cells pump sucrose into sieve tube element; sieve tube element has low water potential; water diffuses in from xylem creating a pressure forcing phloem sap along sieve tubes in leaf; into stem; towards sink, e.g. root; companion cells in the root unload sucrose and may convert it into storage substance, e.g. starch.

Module: Human health and disease

Health and disease (page 58)

1

disease	physical	mental	infectious	non-infectious	degen-erative	deficiency	inherited
measles	✓	✗	✓	✗	✗	✗	✗
cholera	✓	✗	✓	✗	✗	✗	✗
CHD	✓	✗	✗	✓	✓	✗	✗
rickets	✓	✗	✗	✓	✗	✓	✗
lung cancer	✓	✗	✗	✓	✓	✗	✗

disease	physical	mental	infectious	non-infectious	degen-erative	deficiency	inherited
night blindness	✓	✗	✗	✓	(✓)	✓	✗
bronchitis	✓	✗	✗	✓	✓	✗	✗
TB	✓	✗	✓	✗	✗	✗	✗
stroke	✓	(✓)	✗	✓	✓	✗	✗
malaria	✓	✗	✓	✗	✗	✗	✗

Balanced diet (page 60)

1

component	functions
carbohydrates	provide energy
fats	provide energy and essential fatty acids
proteins	provide energy and essential amino acids
vitamins (e.g. A, D)	maintain essential processes e.g. night vision, prevent deficiency diseases
minerals (e.g. iron)	used to make certain compounds, such as iron in haemoglobin
water	solvent; reactant (e.g. in hydrolysis); main component of body fluids such as blood, tissue fluid and lymph; loss in sweat helps to maintain constant temperature
fibre	maintains health of digestive system; prevents constipation

2 Reference Nutrient Intake, Estimated Average Requirement, Lower Reference Nutrient Intake. These are used to assess intakes of groups of people in dietary surveys; used in planning diets for groups of people.

3 Mother's rate of metabolism may increase, so she requires more energy (last third of pregnancy only); materials are required for growth of placenta and fetus. Milk production in mammary tissue requires energy and materials to make fats, proteins and carbohydrates in the milk.

Exercise and fitness (page 68)

1 Oxygen deficit: volume of oxygen required by the body for aerobic respiration at the beginning of exercise that is not supplied by gas exchange and cardiovascular systems. Oxygen debt: volume of oxygen absorbed after exercise has finished above that required for normal activities at rest – needed to respire lactate that builds up as a result of oxygen deficit.

2 Carry out a step test. (See page 69 for details.)

3 Divide some people of the same age into groups. None of them should take regular fitness training. Each group undertakes the same fitness training regime for several weeks at different intensities. Assess improvement in fitness using step test at intervals during training. Record results on a graph. See which group shows fastest improvement.

Smoking and disease (page 70)

1 Gas exchange system: trachea, bronchi and lungs; cardiovascular system: heart and blood vessels.

Cardiovascular diseases (page 72)

1 Cardiovascular disease: any disease of the heart and blood vessels (arteries); coronary heart disease (CHD): disease of the heart; caused by damage to coronary arteries that supply heart muscle with oxygen and nutrients. Stroke: brain damage caused by blocked or burst artery. Atherosclerosis: build up of fatty material in the lining of arteries. This is often the cause of CHD and stroke.

3 Highest prevalence in 'western countries' in northern Europe and in North America (also in eastern Europe). Lowest in Japan. Diets high in *saturated* fat may be a contributory factor to CHD. Saturated fat is converted into cholesterol which is deposited in arteries (atherosclerosis). Diets in the west and in eastern Europe are high in saturated fat; Japanese diet is not.

Infectious diseases (page 74)

3 HIV infects T lymphocytes (see page 78). This gradually decreases the number of these in the body and weakens the immune system. The body becomes less efficient at fighting infectious diseases and preventing cancers from growing. The collection of diseases that result is known as AIDS.

Antibiotics and the control of infectious diseases (page 76)

1 Antibiotics kill bacteria or stop them reproducing. Some interfere with cell wall growth (e.g. penicillin), some interfere with protein synthesis.

2 Antibiotic resistance.

The immune system (page 78)

3

cell	origin	site of maturation	action during immune response
B	bone marrow	bone marrow	divide to form plasma cells, which secrete antibodies
T	bone marrow	thymus gland	stimulate B cells to divide; stimulate phagocytes; kill cells infected with viruses

Immunity and antibodies (page 80)

1 antigen – foreign substance that enters the body stimulating production of antibodies; antibody – protein, made by plasma cells, specific to certain type of antigen; passive immunity – antibodies are not made by the body but acquired from someone else; active immunity – antibodies are made as part of an immune response in the body; natural immunity – immunity gained by either catching an infection (active) or by transfer of antibodies across placenta or in milk (passive); artificial immunity – immunity gained by putting an antigen into the body (vaccination – active) or by injecting antibodies (passive); primary immune response – the first response to an antigen (slow); secondary immune response – any subsequent response to *exactly the same* antigen (faster).

Answers to end-of-module questions

In these answers a semi-colon (;) separates individual marking points. Notice that in many cases there are more marking points available than there are marks for the question. This means that there are several ways in which you can gain full marks for a question.

Biology foundation

1 (a) (i) **A** lysosome; **B** rough endoplasmic reticulum; **C** Golgi body; **D** nucleolus; [4]

   (ii) structures found within a cell; each organelle has a specific function. [2]

   (iii) $\dfrac{87}{1500}$ = 0.058 mm;

   x 1000 = 58 µm; [2]

(b) *leaf palisade has*
chloroplasts; cell wall; large vacuole; [3]

(c) *palisade mesophyll has*
cells of one type carrying out one major function (photosynthesis);

lung has
different tissues carrying out many different functions: e.g. tissue; [2]

(d) see page 5 [3]

(e) thin/flat; short diffusion distance; for oxygen/carbon dioxide; between air/alveolar air space, and blood/capillary; [3]

2 (a)

amylose	✓	✗	✗
amylopectin	✓	✗	✓
cellulose	✗	✓	✗
glycogen	✓	✗	✓

each line correct; [4]

(b) same volume of each juice; add Benedict's solution and boil; add same volume of Benedict's solution; boil for same length of time; deeper colour/more precipitate = more reducing sugar; repeat for reliable results; [4]

3 (a) C, H, O, N; all correct = 2 marks
3 correct = 1 mark [2]

(b) peptide bond; [1]

(c) *fibrous*: collagen; *globular*: haemoglobin [1]

(d) secondary structure; α-helix/β-pleated sheet; stabilised by hydrogen bonds; further folding; gives complex 3D shape; tertiary structure; stabilised by ionic bonds; disulphide bridges; hydrophobic interactions; [4]

(e) *collagen*
amino acids; different monomers; peptide bonds; three polypeptides; (triple) helix; shorter;

cellulose
polysaccharide; β glucose; monomers all the same; glycosidic bonds; straight chain;

polypeptides in collagen held together by hydrogen bonds; covalent bonding between collagen molecules; hydrogen bonding between cellulose molecules; 1 mark for using specialist terms correctly; [7]

4 (a) substrate molecule(s) fit into active site; enzyme–substrate complex; enzyme moulds around substrate(s)/induced fit; substrate molecule put under strain, so breaks to form products; two substrates brought close together so react to form product; [4]

(b) very low activity at low temperatures (e.g. 0–10°C); increasing activity with rise in temperature; optimum temperature; (not always 37°C, depends on enzyme) decreasing activity above optimum; graph to show this with labelled axes; *explanation*: increase in kinetic energy; more collisions between substrate and enzyme; enzyme denatured; irreversible change; loss of tertiary structure; bonds break within molecule; e.g. hydrogen/ ionic bonds; active site changes shape so substrate cannot fit; 1 mark for using specialist terms correctly; [7]

5 (a) nucleotides; [1]

(b) DNA is a double helix, RNA a single polynucleotide; DNA is longer; DNA has bases A, T, C, G; RNA has A, U, C, G; DNA has deoxyribose sugar; RNA has ribose sugar; different types of RNA have different structures, DNA is double helix; [3]

(c)

	pig liver	yeast
A:T	0.99:1	0.95:1
C:G	1:1	1.09:1

1 mark for each. [4]

(d) base pairing in DNA is A–T and C–G; ratios are all very close to 1:1; figures show that quantities of A and T/C and G are the same; hydrogen bonds between bases hold polynucleotides together; DNA is very stable; during replication, polynucleotides act as templates; DNA polymerase matches bases; complementary shapes; ensures replication is exact; new DNA molecules are exact copies of 'parent' molecule; semi-conservative replication; different species have different base sequences; make different proteins; have different genes; 1 mark for using specialist terms correctly; [7]

6 (a) insulin; factor VIII (8); [2]

(b) see page 35 for the stages. 1 mark for using specialist terms correctly. [8]

7 (a) (i) make the spindle; to pull chromosomes/ chromatids, towards poles;

(ii) attaching chromosomes to the spindle; [3]

(b) the daughter cells do not receive the complete set of chromosomes; do not have all the genes; cells cannot function; [2]

(c) one set of chromosomes; half the diploid number; no homologous pairs; [1]

(d) mitosis; haploid cells cannot undergo meiosis; chromosomes cannot form pairs; [2]

(e) one; all cells produced by mitosis will be genetically identical; [2]

Transport

1 (a) distances are too great for diffusion; e.g. from lungs/gut to tissues; oxygen/ nutrients/carbon dioxide; are transported to tissues; quickly/in large quantities; [3]

(b) diffusion; oxygen from air to blood; carbon dioxide from blood to air; down concentration gradients; maintained by flow of blood; breathing to ventilate alveoli; [4]

2 (a) arteries thicker wall; more muscle; narrower lumen; more elastic tissue; no valves; [3]

(b) surrounding muscles contract; squeezes blood; backflow prevented by valves; low pressure in heart 'sucks' blood; [3]

(c) blood pressure; forces small molecules out; through capillary walls; small holes in walls; e.g. water/glucose/ions; [3]

3 (a) graph should resemble Fig. 22.2 on page 47.

(b) (i) increases; figures from graph to show this; [2]

(ii) A; [1]

(iii) during exercise; muscles contract more; increase in respiration; [2]

(iv) Bohr effect; haemoglobin/blood, releases more oxygen; for respiration; maintains aerobic respiration; helps to delay anaerobic respiration; which produces lactate. [5]

(c) placenta is gas exchange surface; maternal blood in placenta; does not have high partial pressure of oxygen; fetal haemoglobin needs higher affinity to absorb oxygen; [3]

4 (a) blood to lungs and rest of body; at different pressures; low pressure for oxygenation in lungs; high pressure for delivering oxygen to tissues; [3]

(b) SAN: pacemaker/sends out impulses to heart muscle; to atria first; AVN: delays impulses; sends impulses to ventricles; Purkyne fibres: transmits impulses to base of ventricles;   [3]

(c) atria contract; atrioventricular valves open; blood forced into ventricles; ventricles contract; blood forced into arteries; aorta, pulmonary artery; atrioventricular valves close; semilunar valves open; ventricles relax; fill with blood; semilunar valves close; valves open and close when pressure of blood is greater on one side than the other; 1 mark for well-written answer;   [8]

5 (a) large surface area inside leaves; mesophyll cells; for gas exchange; absorbing carbon dioxide; for photosynthesis; cell walls are damp; water evaporates;   [4]

(b) xylem vessels; in veins; between mesophyll cells; from cell to cell; cell walls; air spaces; stomata;   [4]

(c) high water potential in xylem; water evaporates from cells; lowers their water potential; water diffuses from xylem; by osmosis; into cells;   [4]

(d) transpiration from leaves; lowers water potential in leaf; water moves up xylem; because of cohesion between water molecules; result of hydrogen bonding; water from bucket moved into stem;   [4]

6 (a) all movement is in the same direction;   [1]

(b) *sink*: place where assimilates are used; e.g. roots for storage; assimilates are transported

in the phloem; *source*: place where assimilates such as sucrose are produced; e.g. leaves;   [4]

(c) end walls have sieve pores to allow sap to flow freely; little cytoplasm to resist flow; plasmodesmata between sieve tube elements and companion cells; to allow movement into and out of phloem;   [4]

7 (a) leaf absorbed $^{14}CO_2$; used in photosynthesis; so sugars became radioactive;   [2]

(b) sucrose made in leaf; from radioactive sugars; transported in phloem; along runner;   [3]

Human health and disease

2 (a) gradual; deterioration/damage; loss in function;   [3]

(b) to make a valid comparison;   [1]

(c) death rates are higher among men; for both diseases; in all countries; highest death rates are for CHD; especially in Finland/UK/US; France and Japan have lowest death rates for CHD; use of figures from table in support;   [4]

(d) the number of people who develop CHD who are (or have been) smokers; the percentage of men and women who smoke;   [2]

(e) raises blood pressure; constricts arteries (e.g. coronary arteries); makes heart work harder; reduces oxygen-carrying capacity of blood; carbon monoxide combines with haemoglobin; reduces supply of oxygen to heart muscle; promotes atherosclerosis; deposits of fat in arteries; promotes blood clotting; leading to thrombosis; if in coronary arteries; leads to heart attack/ angina; 1 mark for using specialist terms correctly.   [8]

3 (a) used to make protein; and other (non-essential) amino acids;   [2]

(b) fat; sucrose; starch;   [3]

(c) *some examples* age; gender; physical activity; pregnancy/lactation; climate (related to heat loss);   [3]

(d) starvation; use of glycogen stores; use of fat stores; muscle wasting; body protein used for energy; emaciated/very thin; poor growth; poor mental development; symptoms of kwashiorkor/marasmus (see page 62); 1 mark for well-written answer.   [7]

4 (a) goblet cells/mucous glands; in trachea/ bronchi; make mucus; sticky; collects dust / bacteria; ciliated epithelium; cilia move mucus up towards throat;   [4]

(b) cartilage keeps airways open; low resistance to flow of air; elastic fibres stretch when breathing in; recoil when breathing out; help to force air out of lungs;   [4]

(c) adrenaline released into blood; *increases in* heart rate; stroke volume; cardiac output; breathing rate; tidal volume; oxygen intake; blood flow through muscles; rate of respiration; blood diverted away from skin/gut; 1 mark for well-written answer. [8]

5 (a) see page 68; [3]

(b) use a step test at intervals; see page 69 for details. [4]

6 (a) unprotected sexual intercourse; via infected needles; from mother to fetus across placenta; in blood/blood products; [3]

(b) problems are outlined on page 77; [4]

(c) *malaria*: mosquito is vector; difficult to control in developing countries; resistant to insecticide. *Plasmodium* is resistant to drugs; no vaccine. *Cholera*: transmitted where limited or no, sewage treatment; poor water treatment; common where refugee camps; poor housing/living conditions;

overcrowding in squalid conditions; poor hygiene; 1 mark for well-written answer. [8]

7 (a) measles virus is an antigen; antibodies have antigen-binding sites; variable region; antibodies are proteins; with amino acid sequence/primary structure; gives tertiary structure; with specific 3D shape; that 'fits' shape of measles virus; [4]

(b) there is an immune response; within the body; B cells divide; to produce plasma cells; secrete antibodies; in passive, antibodies are injected/come from another source; [3]

(c) vaccine contains antigens; put into the body; to stimulate immune response; not 'caught' naturally; [2]

(d) highly infectious; often caught by young children before they are vaccinated; difficult to organise vaccination programmes in developing countries; in cities; where there is overcrowding; and disease is transmitted easily; [4]

INDEX

Note: **bold** page numbers indicate major topics

nucleic acids **28-9**
  as macromolecules 8
  see also DNA; nucleotides; RNA
nucleolus 5
nucleotides 8, 28
  see also DNA; RNA
nucleus 5, 6, 43
nutrition *see* diet and nutrition

obesity 63
organelles 4-5, 24
  chloroplasts 5, 54
  mitochondria 5, 11
  nucleus 5, 6, 43
  ribosomes 5, 32, 33
organs **7**
osmosis 26, 45, 55
osteomalacia 63
oxygen
  deficiency and debt 68
  haemoglobin saturated with 46-7
  production and enzyme 19, 20
  transport 45, 46-7, 48
    diffusion 26, 45, 50, 64
oxyhaemoglobin 46

pandemic 59
passive immunity 80
pathogens 74, 76, 79, 82
  see also antigens
pentose sugar in nucleotides 28
peptide bonds 12
permeability, partial (of cell membranes) 25
pH and enzymes 22-3
phagocytes 44, 78-9
phagocytosis 44, 78
phenylalanine 12
phloem 7, 50
  transport in **54-5**
phosphate
  ion 17
  in nucleotides 28, 29, 30
  in phospholipids 8, 11
phospholipids 8, 10, 11
  in cell membrane 24, 25
photosynthesis 50, 54
physical diseases 58
plaque build-up in arteries 72
plasma 44, 45
  cells 81
plasmid 35
Plasmodium 74, 77, 82-3
platelets 44
polio 82
polymerase 30
polymerisation 8
polynucleotides 32
  nucleic acids as 28, 29
  see also replication of DNA
polypeptides 12
  in antibody 81
  in collagen 15
  in globular and fibrous proteins 14-15
  mRNA translated to make 13, 32, 33
  organisation of 13, 14
population 38
potassium ion 17
potometer 52
pregnancy, diet in 61
pressure points 66
prevalence 63
  of disease 59
prevention of diseases 73, 75, 76-7
primary consumers 38, 39
primary immune response 81, 82, 83
primary structure of polypeptide 13, 14
producers 38, 39
product increased by enzymes 20
prokaryote cells 5
  see also bacteria
prophase in mitosis 36-7
protease 8
proteins **12-13**
  carrier 24, 27, 54
  catalysts *see* enzymes
  in cell membrane 11, 24, 25, 26
  channel 26-7
  in chromosomes 34-5
  in diet 60, 61

energy deficiency 62
globular and fibrous **14-15**
as macromolecules 8
plasma, in body fluids 44
synthesis **32-3**
testing for 13
  see also amino acids; collagen; enzymes;
    haemoglobin
protoctist 6, 74
pulmonary artery and vein 48
pulse rates and blood
  pressure **66-7**
purines 28, 29
pyrimidines 28, 29

quaternary structure of haemoglobin 14

R groups (residual groups) 12, 15, 27
recovery time 67
red blood cells 14, 43, 44, 45, 47
  see also haemoglobin
reducing sugars 9
reference nutrient intake (RNI) 60, 61
rehydration therapy 77
replication of DNA 29, **30-1**
  experimental evidence for 30-1
  mitosis 36-7
  semi-conservative 30
residual groups *see* R groups
resistance of antibiotics 76
resolution 4
respiration 26
  aerobic 68
  anaerobic 68
  and energy flow 39
  and gas transport 45, 46-7
ribose 28, 29
ribosomal RNA (rRNA) 29
ribosomes 5, 32, 33
rickets 63
risk factors and diseases 67, 72, 73
  deficiencies 58, 62-3
RNA (ribonucleic acid) 8, **28-9**
  see also messenger RNA; ribosomal RNA; transfer RNA

RNI (reference nutrient intake) 60, 61
rough endoplasmic reticulum (RER) 5
rRNA *see* ribosomal RNA

saturated fat/fatty acids 10, 72, 73
saturation of haemoglobin with oxygen 46-7
secondary consumers 38, 39
secondary immune response 81
secondary structure of polypeptide 13, 14
self-inflicted diseases 58
  see also smoking
semi-conservative DNA replication 30
serine 33
sieve elements 54-5
sieve tubes 54
sinks 54
sino-atrial node 49
smallpox eradicated 82
smoking and disease 58, **70-1**
smooth endoplasmic reticulum 5
social diseases 58
  see also tuberculosis
sodium 17, 44
sources 54
spindle 37
spirometer 65
squamous cells in blood vessels 43
squamous epithelium 6-7, 65
starch 8
  broken down to maltose 19, 21
  testing for 9
starvation 62
Stentor 6, 42
stomata 51, 53
strokes 59, 72
substrate, enzyme 20
  complex 19
  concentration 22, 23
subtilisin 19
sucrose 8
  in plants 50, 54, 55
  transport of 7
  unloaded at sinks 55
sugars

non-reducing 9
in nucleotides 28, 30
in plants 54, 55
reducing 9
  see also glucose; sucrose
surface areas to volume ratios 42
synthesis, protein **32-3**
systolic blood pressure 66, 67

T cells 74, 78
tar in tobacco 70
TB *see* tuberculosis
telophase in mitosis 36, 37
temperature and transpiration 53
tertiary structure of polypeptide 13, 14
tetanus 80
thymine 28, 29
tissues **6-7**, 78
  in airways 64
  plant 7
tissue fluid **44-5**
trachea 6, 64
transfer RNA (tRNA) 29, 32, 33
translation of mRNA 33
translocation 54
transpiration **52-3**
  environmental factors affecting 53
  inevitable 50-1
  measuring 52
  and movement of water in plants 51
transport
  across membranes 26-7
  in animals **42-9**
  in plants 7, **50-5**
triglycerides 8
  functions of 10-11
trioses 54
tRNA *see* transfer RNA
trophic level 39
trypsin 19
tuberculosis (TB) 58, 59, 74-5
  control of 76, 77, 82
  resistant strains 76
tyrosine 12

unsaturated fatty acids 10
uracil 28, 29

vaccination **82-3**
vacuole 5, 51
valine 33
valves 43
  heart 48, 49
vectors of disease 74
veins 43, 48
vena cava 48
ventilation 64
ventricles 48-9
virus 74
  see also HIV/AIDS
vitamins 60
  A 60, 62-3
  D 63
  deficiencies 62-3

water **16-17**
  addition (hydrolysis) 8, 9
  in body fluids 44
  cohesive molecules of 16-17
  in diet 60
  eliminated (condensation reaction) 8, 9
  as environment 16
  liking *see* hydrophilic
  as liquid 16
  in living organisms 17
  in plants 7, 50-3, 54
    see also transpiration
  potential 26
  properties of 17
  repellent *see* hydrophobic
  vapour loss *see* transpiration
wavelength 4
white blood cells 44, 78

X chromosome 35
xerophthalmia 63
xerophytes 53
xylem 7, 50
  vessels 51